Fred Dibnah

A Tribute

Fred Dibnah
A Tribute

Keith Langston

WHARNCLIFFE
TRANSPORT

First published in Great Britain in 2009 by
WHARNCLIFFE BOOKS
An imprint of
Pen & Sword Books Ltd
47 Church Street
Barnsley
South Yorkshire
S70 2AS

ISBN 978 18456 311 54

Typeset by Mac Style, Beverley, East Yorkshire
Printed and bound in Thailand by Kyodo Nation Printing Services Co., Ltd

Pen & Sword Books Ltd incorporates the Imprints of Pen & Sword Aviation, Pen & Sword Maritime, Pen & Sword Military, Wharncliffe Local History, Pen & Sword Select, Pen & Sword Military Classics, Leo Cooper, Remember When, Seaforth Publishing and Frontline Publishing.

For a complete list of Pen & Sword titles please contact
PEN & SWORD BOOKS LIMITED
47 Church Street, Barnsley, South Yorkshire, S70 2AS, England
E-mail: enquiries@pen-and-sword.co.uk
Website: www.pen-and-sword.co.uk

Contents

Pictured here is proud grandfather Fred with Isobel - the latest addition to the Dibnah family, a cute young lady born to Lorna and Alistair. She arrived in this world on the very day that Fred was in London collecting his MBE (7 July 2004). This photograph was taken two days after he returned from the capital.

Always referred to by the family as 'Uncle', here is Jake Tomlinson seen during the restoration of the convertible.

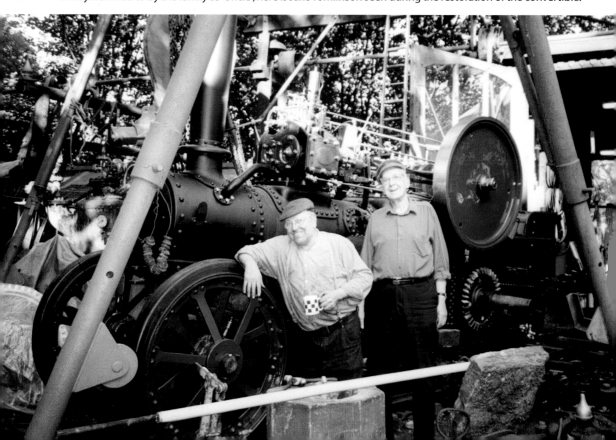

Fred Dibnah
The Family Man

Dr Fred Dibnah, MBE, steeplejack and television presenter, was born on 28 April 1938. He died after a long illness on 6 November 2004, aged sixty-six.

Following his career in television, Fred became a high-profile media personality and the fame that accompanied that status never affected him, or in any way changed his down-to-earth demeanour. He will be remembered not only for his many practical achievements, but also for encouraging thousands of others to care about our industrial heritage. That people from all walks of life have appreciated his efforts is without doubt. He enriched many lives through his work that he always looked on not as a chore, but as a pleasure.

Proof of his many successes is fortunately preserved in tangible artefacts, on film and in the written word. His official recognition was highlighted by the award of the MBE which deservedly came. He had earlier been accorded the tribute of receiving not one, but two honorary degrees. But, moving away from the public side of his life, let us not forget that, first and foremost, Fred was very much a family man.

Born, like a great many of his admirers, in humble circumstances, he was brought up to appreciate the values of companionship, honesty and the satisfaction to be gained from a job well done. As readers digest the comments of his friends and colleagues in this all-too-brief compliment to him, they will see that the one word which constantly comes to

1991. Fred holding Jack, while Sue holds 'baby' Roger.

Lorna, Jack, Jane, Caroline and Roger.

Lorna, Fred and Jane.

Lorna and Roger at a chimney topple.

A very young Jack on the living van steps.

mind is love. He was a greatly loved human being, and he left his mark on the world he lived in and accordingly it is a better place for his having been there.

Thanks are due to the many people who have unselfishly passed on accounts of their involvement with Fred so that others can enjoy the retelling of them.

Fred is survived by his brother, Graham, his widow Sheila and his ex wives Alison and Sue.

He was the proud father of five children: to Alison there are daughters Jane, Lorna and Caroline, and to Sue there are sons Jack and Roger. With his last wife, Sheila, Fred took on the mantle of stepfather to her son, Nathan. He was a loving grandfather to Jane's sons, Christopher and Daniel, and also to the newest addition to the family, Isobel, who was born to Lorna the day he received his MBE.

Fred's family have been kind enough to allow us an insight into their private world by agreeing to a collection of their personal photographic memories being included in this tribute. Heartfelt thanks are due to them, so let us remember that we are in a privileged position and, as such, we should enjoy the images and thereafter respect the family's privacy.

Roger pulling the chains that work the bosun's chair.

Jack with a home-made engine (buggy) at the family home in Radcliffe Road, Bolton.

Left: Jack and Fred share a moment on location.

Below: Relaxing with Jack and Roger after filming in London.

Fred and Sheila at Buckingham Palace after receiving his MBE.

Steeplejack!

1993. Roger helps with cleaning the roller wheels.

Never one to be afraid of heights, Fred plays Father Christmas on the chimney of his mother's house in 1997.

Roger in the snow at Radcliffe Road.

Caroline, Alison, Lorna and Jayne.

Lorna.

Fred's coming-of-age announcement in 1959, placed in the local paper by his parents.

Master of art. Fred's artistic talents on a letter addressed to his mother, whilst serving in Germany.

Above: Christmas Day 1995. Fred playing with a set of terracotta bricks.

Right: The Dibnah boys in London – Roger, Fred and Jack.

Looking Back
The Early Days

Fred was a Red Rose man through and through – but he could so easily have been born a White Rose man! By way of explanation for those domiciled on the other side of the 'English Alps', we should point out that there is a strong Dibnah connection to savour as Fred's father, Frank Dibnah, was born in Hull, East Yorkshire. It was Frank's father, Alfred (Fred's grandfather), who first took the Dibnah name over the Pennines. Like a good many in those days, he had to travel from his native home in order to find work.

He and his young bride Hannah (nee Ford) were just starting a family. He was a newly trained locksmith who chose to settle with his young family in the Lancashire mill town. Having qualified as a master whitesmith, in 1903 he successfully secured employment there and started working at the Bolton company of Fisher Raworth Ltd, which was then situated on Bradshaw Gate. Trading in architectural ironwork, the company was founded in 1811 and is still going strong today. The 185-year-old company, however, now trades from premises in Spa Road, Bolton.

We spoke to Elizabeth O'Neill, present-day company director of Fisher Raworth, who said that she was both surprised and pleased to hear of her firm's Dibnah connection!

At school in Bolton, young Frank Dibnah was in the same class as a young lady who grew up to be one of the UK's best-loved actresses, none other than 'She Knows, You Know' Hilda Baker!

However, the Dibnah 'White Rose' connection does not end there as Fred's nephew, Carl, the son of his only brother, Graham, has crossed the 'Alps' in the opposite direction to his great-grandfather and now lives and works in Hull.

Visiting Fred's home town of Bolton as an act of remembrance some twelve months after his passing, we were privileged to be invited to the home of Graham and his wife, Audrey. Graham is a builder and he and Audrey have a grown-up family comprising a son, daughter and two grandchildren.

We are grateful to Graham and Audrey Dibnah for recounting their personal reminiscences, which serve perfectly to afford us a fascinating glimpse into the life of the young Fred Dibnah.

My big brother
Graham Dibnah fondly remembers Fred

Graham was born in 1942, the younger of the late Mr and Mrs Dibnah's boys. Frank and his wife, Betsy, brought up their young family in a neat terraced house only a stone's throw from the once-famous Burnden Park football ground, the former home of Bolton Wanderers FC, who now ply their Premiership League skills at the futuristic Reebok Stadium on the town's outskirts.

As all Fred's admirers know, he was not in any way a follower of the so-called beautiful game or, indeed, any other sporting activity that didn't involve a set of spanners. Nevertheless, the old stadium was to be a part of his young life.

Graham recalls that Fred did express a wish to go with him to a cup tie at Bury but it transpired that he only wanted to do so in order to travel on a train over the viaduct at the back of their house. When they got to Bury, Fred simply sat on the station platform, presumably train-spotting all afternoon, and then returned on the soccer special with his father and brother after the match.

All young boys at that time needed to secure an income, not only for their own

Fred Dibnah talks to an interested spectator after felling the chimney at Ocean Mill, Great Lever, Bolton. 3 October 1982.

The River Irwell just after it joins the River Croal, in the topright hand of this picture is the embankment up which Fred and 'Heapy' pulled the canoe. The canal used to run just in front of the house at the top of the embankment and a grit stone retaining wall can still be seen.

needs but often to help eke out the family budget. For the young Graham Dibnah, the close proximity of the Wanderers' ground was doubly significant. Firstly, he was then, and still is, a keen supporter of his home-town team. Secondly, the local boys derived a spot of ad hoc pocket money from Bolton Wanderers that was, as Graham recalled, rather dependant on how well the team was doing at the time.

In those days it was the custom of the football club to offer for sale tickets for their next home match and sometimes, if a cup run was in progress, away tickets as well, on a Sunday morning. There were strict rules regarding the sale of those valuable bits of paper, the important one, as far as the local kids were concerned, being the one that said that there would be only one ticket sold per person.

Graham and his pals would therefore take a place in the queue for a supporter who needed more than one ticket and then buy a ticket, with the supporter's money, on his behalf. For that service they would receive a sixpence or, if they were really lucky, a whole shilling! By lining up sometimes with a cap on and then without one, they could often go around twice, 'unclocked' by the turnstile men. In most cases Graham would spend his earnings on tickets to watch the football himself. It is perhaps no surprise that Fred did not participate in these particular money-making activities, always preferring instead to go adventuring!

The Dibnah family were, like all Bolton folk, deeply touched by the events at Burnden Park on 9 March 1946 and were lucky not to have lost family members on that fateful day. Graham recalls that their father, Frank, was himself a turnstile attendant and that grandmother Rachael (on his mum's side) was a keen Wanderers' fan. She was on the terraces and Frank on duty at the time, thus two Dibnah family members could very easily have lost their lives in the Burnden Park disaster.

The tragedy occurred during the FA Cup encounter with Stoke City FC, whose team on the day included the great Sir Stanley Matthews. The official attendance given at the time was 65,419 but an estimated 85,000 people poured into the soccer ground. The grandstand collapsed, claiming the lives of thirty-three supporters. The site is now a supermarket and a memorial plaque commemorates the sad event.

Graham says his dad heard a distant rumble and the sound of shouting, neither of which was obvious as being from the either nearby railway line or associated with normal football ground banter.

Somebody shouted to him: 'Frank, bolt that door.'

'Bugger that,' he said. 'I'm off!'

He ran into the street in time to see rubble dislodged from the collapsed stand, falling at the back of the kiosk where he'd been on duty. Grandma Rachael told them she and her friends ran from their places on the left of the collapsing stand and, like thousands more, found safety on the playing pitch. It was, she'd told the boys, a frightening and bewildering experience and it was several hours before she and her family were reunited.

Fred did visit that fateful football ground. At the beginning of his steeplejacking career, he found employment climbing the floodlight pylons in order to work on the fittings. Graham observed that this was nothing new to Fred as he'd climbed the lights often before for a schoolboy dare – of course, unbeknown to his parents.

The brothers attended St Michael's Primary School. From there Fred went on to become an above-average pupil at Bolton School of Art. Graham, who completed his education at Heywood Secondary Modern School, went on to be apprenticed to a bricklayer and still works in the building trade.

Boy racers!

Fred was not the greatest sports fan – his interests lay in a completely different direction. He was, though, a very accomplished model-maker and Graham recalled that, among

other things, he built a fine scale model of *Titanic* that, he says, was almost 4ft long and over a foot high. The school encouraged Fred to enter the model in a competition for schoolboys in Lancashire and Cheshire. He did just that and won hands-down!

Some of his schoolboy projects were not so successful. Graham recalls that a racing buggy he built in partnership with his regular cohort, Alan Heap, was such an example.

A very young Fred high above Bolton circa 1958. In the right foreground is Burnden Park, the then home of Bolton Wanderers FC. Behind the ladder is a steam train heading for the town's station.

Fred Dibnah and wife second wife, Sue, with their son Jack, aged three, in February 1991.

It was, without any doubt, a good runner but Graham remembers that it lacked a lot in the control department, to say the least. It was constructed using a plank, a tea chest and four pram wheels rescued from the canal (more of Fred's canal exploits later). A street near where Fred lived had an attractive downward racing slope, culminating in a junction with the main road, ideal for schoolboy would-be speed record-breaker.

After many low-speed trials during which time Fred tested his patent braking system thoroughly, he and Alan decided to go for broke with a run down the slope, effecting a left-hand speed turn at the bottom.

Graham recalled standing at the top, watching as Heapy (who gave the push start) jumped on board. The machine really took off! It was apparent that the steering worked perfectly – but the same could not be said of the brakes…

Freds brother Graham in November 2005.

As the two speedsters reached the junction of Western Street and Manchester Road, what should appear, slowly crossing from right to left, but a No 8 Manchester-Bolton bus! Fred first swerved to the left but presumably seeing hordes of people waiting at the bus stop, was left with no option but to resume his straight-on course and brake violently.

'Break' is a more appropriate word than you may think, as the brake gear simply followed the rapid spinning movement of the rear wheel it was intended to slow. Graham observed that it flew off, scattering its component parts about the cobbled street.

Now the Manchester Corporation Transport Leyland double-decker bus of those times was a very solid object indeed, and the boys were lucky to escape with only slight injuries as they crashed full-tilt into the rear of the stopping vehicle.

The bus conductor, a man Graham described as huge, sprung from the platform and, in the custom of the time, grabbed both boys by the scruffs of their necks, and marched them to Fred's front door.

A 'grass' in the gathering crowd had indicated the home of 'that 'un', an indictment he loudly proffered while pointing to a very shaken Fred. It was early to bed, no tea and banned from playing out for a week. Being confined to barracks did not faze Fred – at least not until his later Army days – and he used the time in his bedroom to work on another of his many cunning plans.

Any old iron and matters aquatic!

Graham Dibnah, Alan Heap, Alan Williamson or, indeed, any of his school friends will tell you that, in addition to the bleach works and the BR engine shed at Bolton (an ex-Lancashire & Yorkshire Railway depot, coded by BR 26C from 1948-1963 and thereafter 9K until the end of steam in 1968), the canal and nearby river held a great fascination for Fred.

Those two watercourses were repositories for a great many discarded bicycles, prams, lawnmowers and other such paraphernalia, all of which were seen by Fred as treasures beyond the dreams of avarice. The Manchester, Bolton and Bury canal was then in an abandoned state and the river in question was the Croal, which connected with the River Irwell.

Walking the banks of these watercourses had kept Fred happy for a few years and he and his pals collected what he described as 'loads of plunder' – any old junk, by another name. This he stored in his parents' backyard and even in his bedroom. 'You never know when it will come in handy,' he used to say.

With growing up came a greater thirst for adventure and our hero told his younger brother that he needed to get both into and on the water. He reasoned that dredging for

October 1964 Fred, his father Frank, his mother Betsy and brother Graham at Graham's marriage to Audrey.

Looking down from the work platform at the top of the 275ft high Edbro Ocean Mill, a dramatic sight. No wonder Fred's mother didn't want him to climb!

treasure with a rope and steel hook was only half the answer. More drastic action was needed.

As for the 'into' bit, the long, six-week school summer holidays often provided warm days so the lads had no fear on those occasions of stripping to their underpants and getting into the water in search of even more plunder, which they reasoned must lie in the murky depths. This was OK but not really enough for our Fred, as the highly-coloured water of the cut was hard to see into. A solution to that problem was proposed.

Having seen detailed pictures of a deep sea diver in a copy of *National Geographic* (see – young boys did look at other things, too!) he decided to build a diving helmet, although what he really would have liked, as he told Graham at the time, was a diving bell or, better still, a full suit!

Using various items including bits of oil drum, a biscuit tin, some rubber hose, various pipe fittings, hemp rope and waterproofed oilcloth, they fashioned a diving helmet. Graham is not sure just how well this equipment worked, or, perhaps, didn't, but suffice it to say that Fred and his pal didn't drown.

After putting together this Heath Robinson contraption, Fred decided to test it out before

Up up and away in 1959 Fred starts to climb a very high chimney with his then newly acquired ladders.

Danger, man at work. Fred using a bosuns chair to move up and down a local church he was working on.

venturing into the muddy waters of the Bolton Bury Canal. The bath at home was, for security reasons (mum on the prowl) a non-starter although he did secretly test that it was watertight by ducking it under a couple of times. To test it fully he decided that, with Heapy, he would trial the helmet in the tranquil waters of the local swimming baths.

Fred did not swim and is on record in later life as saying that 'the only thing you should do with water is make beer from it or put it in steam boilers', but that did not put him off.

However, the manager of the baths did. He spotted the two pals trying to smuggle the home-made, Jacques Cousteau type diving equipment into the pool and sternly ordered them out.

Graham and his friends looked forward to all-day football games during those fondly remembered and seemingly always gloriously sunny days. Fred and Heapy would also be up and at it early but in their cases, it would be to prepare for some adventure or other, usually involving water and almost always being an activity tinged with more than a hint of danger.

The same magazine that inspired the diving helmet was probably responsible for giving Fred the idea and encouragement to build a boat. It was a canoe built from 2x2in timber framing with an old wagon tarp (tarpaulin sheet) stretched over it tightly and nailed securely in place. Graham remembers that the canoe was built to a very good standard indeed and that, in particular, the joinery was of the highest order.

Even then, at the age of about fourteen, Fred showed a real aptitude for working materials with his hands. After he'd put on the skin (tarp), he gave the whole thing a coat of paint. Making the necessary paddles was, for a guy of Fred's ability, child's play. There was, however,

Fred's sign on the 275 Ocean Mill chimney. Note the Burnden Park address.

Fred in mid climb.

one problem: Fred had quietly built the canoe in his bedroom with Graham sworn to secrecy. By fair means or foul, he had craftily managed to keep his mum out of the room while construction took place.

However, the good ship 'Dibnah' could not be extricated from his bedroom in its finished state. Fred was unable to negotiate the landing of the terraced house with the generously sized, two-seater open-topped boat. He tried the manoeuvre unsuccessfully several times while his mum and dad were both at work. Launch day was put on a back-burner. Fred and Heapy went into a huddle out of earshot. The plan that followed later was, to some eyes, mad in the extreme but to Fred, merely the application of pure logic!

Picture the scene as described by Graham, who had not an inkling of what Fred had in mind by way of a solution: All the family is at home on a quiet Sunday morning and mum is at the sink preparing the vegetables for the roast. Dad is reading Saturday night's football 'pink' and Graham is in the back entry kicking a ragged 'case and tube' against the wall, waiting for his mates as they had arranged to recreate an England v Scotland soccer encounter that England, of course, would win. Fred is nowhere to be seen, but his mum remarked that she could hear him upstairs.

'Strange,' thought Graham. Heapy was wandering around at the front of the house and, with hindsight, Graham recalled that he looked furtive. He had been in the house earlier and had taken a bag of some sort up to Fred's room.

But the time for the big match was looming and Graham dismissed all other thought from his mind, except, that was, for the need to nip indoors and fill up a bottle with cold water. He recalls that mum was still working at the sink and, when asked, started to fill the bottle for him. It was then that all hell seemed to break loose!

Fred and workmate Eddie Chatwood reach the work staging at the top of a very tall chimney.

At primary school together, Graham aged seven (left) and Fred aged ten.

Growing up, Fred at eleven (left) and Graham aged eight.

Fred aged five, Graham aged just one and the boys cousin Audrey who is regaled in her May Queen dress in 1943.

Graham is checking that his best man Fred has the wedding ring intended for his bride.

'What the blazes is all that banging upstairs?' she asked her husband. Mind, they were well used to their would-be inventor son doing things that were out of the ordinary.

Frank Dibnah started to say 'I don't know', when Graham, who had just left by the back door, ran back in through the front, in a state of highly animated excitement. Following him they all piled outside to see that the bedroom window had been taken out (so that's what was in that bag – tools, thought Graham). Fred was leaning out and Heapy was below taking delivery, it seemed, of the newly constructed canoe.

Eventually calm was restored and Fred put the sash window back into its frame without disturbing a single speck of paint, thus saving his skin. The launching was rescheduled and the canoe worked a treat, so much so that the boys sailed the rapids of the-then swollen River Croal and even ventured on into the River Irwell.

The intrepid pair then dragged the boat up a very steep embankment so that they could then navigate a section of the canal.

It was gone nine o'clock that night when Graham, tucked up in bed, heard his big brother trying to put the boat in the backyard quietly and sneak into the house. Fred's parents had been frantically trying to find out where he was and mum had just put on her coat in order to go and inform the police that the boys were missing. Fred just wandered in as if nothing had happened, declaring that the boat was a huge success and the voyage a great adventure. No supper for him that night!

A shocking discovery!
Those local watercourses had another attraction to Fred. They were ideal places for target practice, whether that be with the catapult or, in his later teens, the gun. Rats abounded in those dank urban waterways and, being vermin, they were fair game to Fred and his mates, but on one such bank-walking expedition Fred, Graham and Heapy got far more than they bargained for.

It was, recalled Graham, a weekend jaunt and it started out as a search for bike wheels and also a rodent hunt.

The boys had ventured to the point where the River Croal joined the River Tong and were actually quite near to Fred's later home in Radcliffe Road. The balmy summer's day was tailor-made for the Gang of Three and they had already successfully loosed off a veritable battery of bricks against a family of rats seen skulking around a riverside culvert when Fred spied a set of half-submerged pram wheels that, he declared, warranted further investigation.

The boys had recovery off to a fine art. They were used to utilising a home-made hook, a length of rope and any available tree branch to claim their bounty. If the weather was warm enough, one of them – usually Fred – would even get into the water while the others spasmodically hurled bricks in the direction of the rodents in order to keep them at bay.

As they trawled about with a branch in the rain-swollen river they latched onto the wheels and began to pull them, a movement, it transpired, which was to dislodge something else.

Fred was the first to spot it.

"Look," he cried. "There's a boot in the water."

Indeed there was – but the problem was that it was attached to a foot…

As the lads cleared away the debris that the rising water had forced against a sunken pipe, they saw that the foot was, in turn, attached to a body. As the bloated torso

Treasure hunting on the Bolton & Bury Canal – Dibnah and 'Heapy' style.

Fred and an invention he called 'Spiny Norman' (AKA The Sputnik) which was a clever device for moving the scaffolding work platform down a chimney as physical demolition was under way.

Fred starting to climb a big un!

emerged, it was clear to see that it was sadly the earthly remains of a lady, even though Graham recalls that there was a great deal of rodent damage.

The boys set off running to the nearest location where they thought they could get help. That was at the factory of Roberts Croupline Ltd, the famous cough medicine manufacturers.

The gateman at the works phoned the police and he sent the boys back with instructions that they should stay with their gruesome find until the bobbies arrived.

When they turned up half-an-hour later, the first thing they did was to tell Graham, presumably because he was the youngest, to "Go home and tell your parents that your brother and this other lad are helping us with our enquiries."

It transpired that the unfortunate lady had gone missing some weeks earlier. The authorities thought that she'd fallen into the river, either by accident or design, and had met an untimely end. Following a sharp summer shower, floodwater had caused her body to float free from where it was previously trapped and so it became visible to our intrepid explorers. Some days later, the police visited Bolton School of Art in order to interview Fred further which caused quite a stir among the other pupils.

The clock stopped there!

Fred's ability at joinery had been noticed when he was at college and the teacher in charge of cabinet-making encouraged him to make a casing for a grandfather clock, which he did to great effect.

He then went off to do his two years' national service, which for those of his intake was then extended to two-and-a-half years. Some weeks after he arrived home from serving Queen and country in Germany, the man from the college came to the house and asked Fred if he would kindly remove the aforementioned item from the joinery classroom where it had languished since his conscription.

It stayed at the bottom of the stairs in the family home for a very long time before Fred eventually got around to putting a face and mechanism into the superbly-made case, said Graham. When Fred married and left home, the clock went with him.

Graham recalled: 'It was really well made and a real tribute to our kid's skill with woodworking tools. From looking at it you could see that he had an artist's eye.'

Planning for the future!

When Fred left school he was, as has been well documented, only interested in one type of work and that was steeplejacking, although his friends and teachers at the art college will all confirm he could easily have made a living with a pencil as a draughtsman or a set of brushes as a commercial artist.

During his last summer holiday from school, he spent day after day watching steeplejacks at work on two chimneys at the Roberts Croupline factory and, though he certainly loved drawing, his real heroes were those tough guys high up on the staging.

He secretly started to buy ladders and, as his mother was definitely not keen on him climbing, he hid them. Unbeknown to the family he had already started doing such work and was going out pointing houses, hence earning the cash to buy his tools of the trade. To climb his beloved chimneys he would need many sections of ladder and so the surreptitious work went on for quite a while.

But where do you hide sets of heavy, long ladders? Again Fred's ingenuity came into play.

Fred's Dad Frank looks on in surprise as Fred and his adventurer pal get the canoe out of the house via Fred's bedroom window.

Along the street lived a painter and decorator called Tommy Quinn who sometimes went for a pint with Fred's dad, Frank. One day when he knew that his dad was at work, Fred went along to the King Billy public house and engaged the painter in conversation after investing in a pint for the man. The ploy paid off and Tommy agreed to let Fred store his ladders in the yard, promising not to tell his dad. After all, where better to hide ladders than in a painter's store yard?

As Fred's stock of ladders started to grow he had secretly started to climb for real money. Graham recalls that early one morning while in the kitchen with his mum she happened to look out of the back window. Seeing two guys working at the top of the chimney at Entwistles mill in Western Street, she declared: 'Look at those crazy beggars – they must be mad!'

'Little did she know that those "crazy beggars" were, in fact, our Fred and his friend Alan Heap,' said Graham. 'I grabbed my dinner bag and headed off to work. No way was I going to tell her and neither did I want to be there when she found out for herself!'

Graham did work for a short while with Fred, specifically during the harsh winter of 1963. When he was laid off from the building trade because of the severe icy conditions, he was able to earn money pointing chimneys with his big brother. Graham got married

An early colour photograph of Fred's roller Betsy, on the move around Bolton.

before his brother and, on that lovely day in 1964, Fred was the best man at his wedding to Audrey.

She recalled: 'He was the perfect best man and, as you would expect, he was also the life and soul of the party.'

Runaway romance!

The next incident Graham recalled was also to do with marriage but concerned Fred more directly as he was to be the groom.

Furthermore, the unexpected wedding was not to take place in Bolton but over the border in that Scottish town beloved of all romantics, Gretna Green!

Graham takes up the story: 'It was very unusual for mum to come to our places of work so when the foreman called me to the site office I was flabbergasted to see her sat there holding a brew of tea which the guy had kindly made for her and looking more than a little tearful.

'She told me in quiet tone, between the occasional sobs, what she had come to my work place for.

'She was talking very quietly as, in her way, she did not want to wash her linen in public, as we say in these parts, and some of the other lads were in the cabin as it was morning brew-time. She had taken more than a five-mile bus ride to get from our home to the site in Farnworth and, having also walked a fair distance, she needed time to catch her breath.

'When she finally got herself sorted out and calmed down, she gave me news which almost knocked me sideways! 'Your brother has run away to Scotland and he says he is going to get married at Gretna Green to his sweetheart, Alison.'

'I had been used to Fred pulling a few strokes in the past but this one took the biscuit. We were all very shocked and surprised – especially my mum! But that was Fred and he really never stopped surprising people right up to his dying day.

'He was as good a brother as a man could have, and I am very proud of what he became.'

The view from above

Fred always made it known that he spotted Alison, the young lady who he married at Gretna, from the top of a chimney.

He used to say there was no better place to see the comings and goings of the world than from the vantage point offered by an old chimney. He may well be right! The majority of folk only share such vistas occasionally and then only courtesy of an aeroplane or a multi-storey hotel.

As for aeroplanes, Fred was, to put it mildly, not a great lover although he did once fly in a hot air balloon – however that's another story.

In addition to the vistas which he had spied from one such lofty perch, he also spotted the comings and goings of a young lady whom he had watched not only going to school but later, going to and from her place of work.

Eventually Fred found the courage to talk to her. A courtship began and after knowing her for only six weeks or so he took her on a very special shopping trip. Much to her surprise the specific item to be purchased was an engagement ring.

Although she did not know exactly what Fred the romantic had in mind on that day out in Manchester, she was soon to find out! As the man in the shop Fred had ushered her into opened a large tray of rings to show her, she began to get the message.

Fred, in his own inimitable way, then popped the question.

'Pick one – we are getting engaged tonight,' he said. She did just that and their runaway wedding date was set for later that spring.

The romance, the seventeen-year marriage and family life of the pair has been well chronicled by others and I only briefly mention it here in order to introduce another side of Fred's complex make up to the reader, that being, of course, the romantic side.

In addition to Fred's many great qualities, his skill as a communicator, his strength of character and his obvious talent for understanding all matters concerned with engineering, there were many other facets to his fascinating personality. Romance was one of them – after all, the number of people who can say that they ran away to Gretna Green to get married are very much in the minority!

Fred liked to tell the tale afterwards, always emphasising that he pointed the vicar's gable end while waiting the then statutory twenty-one-day residential period before their wedding.

Interestingly, he also pointed the gable end of the small private hotel they stayed in. He said that he did it out of boredom, but really he was hoping to get something knocked off the bill. That did not happen, as Fred often recalled.

'The owner of the digs was from Yorkshire and believe me, in true Yorkshire tradition, he didn't believe in giving owt away!'

Of course, as some of Fred's close friends often reminded him, his own grandfather was from that fair county!

At Home with Fred Dibnah

In the early spring of 2004, *Old Glory* magazine decided to talk to Fred about his then-forthcoming TV series *Made In Britain* as many of that journal's readers had been approached with a view to being included in the filming sequences. In order to set the scene, Fred was interviewed during a series of visits to his Bolton home.

'At home with Fred Dibnah' proved to be one of the most popular items ever carried by the magazine and was something Fred himself said he enjoyed reading.

Fred's enthusiasm for the publication had grown over the years and therefore it was no surprise when he asked one of its writers and photographers, Keith Langston, to cover the story of the filming. But first, the story of those visits to Bolton.

History gave us some great British engineers who fascinated the whole of the developing world with their ingenuity. Modern generations have been able to 'revisit' those engineering workshops of the past through the magic of television and so can

Fred proudly shows off a home-made hydraulic riveter at his home workshop in Bolton in the summer of 2004.

Fred working on the mineshaft.

Fred uses his superb drawing skills to illustrate the details of the mineshaft.

Building the winding house roof.

appreciate the achievements of those great artisans and entrepreneurs. This good fortune is enhanced by the unique descriptive skills and oratory style of a man who, if born in a previous age, would have striven to have been a contemporary of Brunel, Stephenson, Arkwright and the like. Fred Dibnah is that man.

With Fred, what you saw was very definitely what you got. Some say, and not without a little humour, that you would never catch him with clean hands. That does sum up this proud son of Bolton who would never ask anyone to do anything he wouldn't tackle himself. The great beauty of Fred's love of engineering and, in particular, all matters steam, was that he was not content merely to explain how something works - he needed to be sure that you understood the 'nuts and bolts' as well.

The wonderful gift he had was embodied in that God-given skill which all great teachers possess. He would read the understanding, or lack of it, in the recipient's eyes, and thereafter tailored his discourse to a level of complication his senses told him the listener was able to comprehend. He so refined this knack that he was even able to communicate in this manner through a TV camera, as those devotees of his many programmes will be happy to confirm.

You got the feeling, listening to Fred, that your understanding of the subject was what really mattered to him, and all his descriptive powers were focused on achieving that end.

Fred checking the quality of the paint finish on the Aveling's side tank.

It was a sharing thing. He knew how, and that knowledge had given him satisfaction. He wanted you to share that understanding and therefore he was offering you something he saw as pleasure.

The word 'raconteur' can confidently be used to describe the story-telling eloquence and expertise of Mr Fred Dibnah. Storytellers since 'Adam was a lad' have been our greatest link with the past and it is through them that so much of our history has been passed down. Scoff not at the art of slight exaggeration or verbal embellishment but consider whether, without their employment, you would have taken in the details of that all-important, but perhaps otherwise boring, story.

A piece of chalk, a pencil and a length of stick against the soft earth have all been the tools of the natural orator over the years. Add to these an adjustable spanner and a home-made hydraulic riveter and you have Fred Dibnah.

What, you might ask, started him on this strange but wonderful life of danger and discovery? How could a joiner from post-war Lancashire become a media icon, with the advent of satellite TV on a global basis, and yet still be 'Na then, Fred' to so many on a daily basis?

A writer colleague told me: 'Leave plenty of time when you call on Fred. It's a kind of manic open house where engineering problems are related and solved on a relentless

An early penchant for tall towers? Fred the Army cook building a stack of beer cans in the Naafi.

basis; and they take precedence over all else. Except, that is, when midday approaches and everything really does stop for the pies.'

He was absolutely right and, furthermore, at noon, without ceremony, the 'boss', having been told that it was his turn, wandered off to make the tea.

In the moments that followed, victuals enjoyed and a mug in hand, the storyteller qualities in Fred seemed to reach new levels and he needed no encouragement to relate the details of his current crusade – 'the matter of the mine shaft in the back yard' and what he considered to be the metropolitan borough council's 'strange attitude' towards it. But more of that later.

The steam bug infected Fred at a very early age following illicit visits to his father's place of work, a bleach factory. Everyone in those days, he recalled, took their main holidays during the traditional Wakes period and that meant factory shutdowns and time off work for most, but not for his father. Mr Dibnah Snr was a labourer and, as such, was often engaged in that all-important part of factory life – 'shut-down maintenance'.

He would leave a fencing plank pulled loose so that young Fred could gain entry to the works and join him among all the steam contraptions, pulleys and belts. To a young man with the enquiring mind of Fred, the bleach factory was an absolute Aladdin's cave and viewing it sowed the seeds in that youthful mind that were to flourish in adulthood and bring forth the love he held for steam power.

One of Fred's greatest regrets was that his native Bolton is no longer rich in engineering firms producing and repairing machinery, but instead he proffered: 'Bolton is now the

Almost ready for the road, Fred stands proudly with the Aveling & Porter convertible engine – twenty-seven years in the restoration.

world leader in the production of disposable bedpans and paper products closely associated with the disposal system of the human anatomy.

Times do, of course, have to change, but will we, as a nation, live to regret the loss of all those great engineering skills? We are still breeding fertile minds but what are we filling these young heads with now, compared to the skills of the past our forefathers were able to offer us? It may have been said before, but the discontinuation of the indentured apprentice schemes has a lot to answer for. Don't believe me? Then try getting hold of a time-served plumber when you need one.'

Talking of matters educational, it is worthy of note at this point to recall that Fred was the recipient of two honorary doctorates – one from Birmingham University and the other

Fred the teacher, explaining how the throatplate for the Aveling & Porter traction engine was made.

bestowed by the Robert Gordon University of Aberdeen. In both instances, the recognitions were accorded by the relevant engineering faculties but, in typical manner, Fred preferred to recount that they were, in his opinion, given for what he called 'back street mechanic-ing'.

Fred's schooling at St Michael's, Great Lever, saw an early indication of his interest in engineering, when he won a prize, aged 11, for building a model steamship. This was followed by a secondary education that included a session at Bolton College of Art – hence the honing of his self-evident and natural draughtsman skills.

Among his contemporaries at that seat of learning was a young man named Donald Jackson, who is now Queen Elizabeth II's official 'scribe'.

Fred was set to the tools and thus the finer points of 'wood butchery' were taught to him, a trade his father was sure would see his son earn good money and gain security. While pursuing that craft, Fred became fascinated with high buildings and, in particular, the techniques of steeplejacking began to interest him.

Away from workaday joinery, he began to earn extra cash performing 'foreigners' at the weekends. Those extra tasks included the maintenance of weathervanes and the repairing of lightning conductors on chimneys. Fred was not yet scaling the real monsters that were to become his eventual trademarks, but structures of a more modest nature which, he recalled, were 'only' around 100ft high!

This extra income included the then-huge sum of £60, earned for pointing a local mill owner's house. That imbursement led to the burgeoning Dibnah business empire purchasing its first mode of transport – a 1927 AJS motorcycle which, after a degree of restoration, became Fred's pride and joy.

Standing in the way of his business expansion plans were his youthful looks, as he put it. 'No one would, on seeing this young fellow at their door, think of him as being capable of climbing the chimney, let alone repairing it.'

Friendship was to be the architect of his salvation and his old art school master, Lonsdale Bonner, was to come to his aid.

He 'happened upon' Fred cleaning his ladders one day. 'What's all this, Dibnah?' the teacher asked.

'It's a steeplejack business I'm starting up,' Fred told him. Fred went on to explain the problems of his youth getting in the way of obtaining work.

The teacher said: 'Tell you what, I will go and see some people for you.'

So it was he who initially knocked on the doors and got the work. For that invaluable service Fred paid him a tenner a job. By the time the customer saw the young steeplejack it was too late to turn back. He was up the ladder with the work well under way. 'It was a great scheme,' Fred added with a wry grin, 'Lonsdale could get where castor oil couldn't; we had a lot more work in no time at all.'

While his TV programmes have latterly focused on the industrial heritage side of Fred's life, it should, of course, be remembered that, for a long time, he earned his living climbing things – usually very tall things. It was that work which first brought him to the attention of the BBC but first he was summoned, as a great many were in those times, to lend a hand elsewhere.

Fred was called on to serve Crown and country for a couple of years. Another institution of the past which some believe would be the panacea for all ills if reinstated, National Service was to request the pleasure of his company. Fred was, as far as he said he could remember,

not responsible for poisoning anyone but his spell as an Army cook, coming as it did after the statutory spell of 'square-bashing', thankfully did not last very long.

Germany, or to be precise the then Army of Occupation in West Germany, was to be the extent of this 'Join the Army, see the world' young man's travels but, in a phrase much used today by those of a certain political leaning, 'things could only get better'. 'Soldiers,' said Fred, 'did then, as now, the soldiering, the officers shouted a lot and, in general (no pun intended), seemed to live better.'

Part of living better, at least on the camp Fred was posted to, involved riding horses and even, as it turned out, 'riding to hounds'. That, he recalled, proved to be an excellent thing for furthering his Army career. There was an old farmhouse with an outbuilding on the outskirts of the camp and it looked as though the 'heavy mob' had used it for grenade practice on a regular basis. It needed rebuilding, and Fred was to be one of the conscripted workers.

A high ranker said: 'You're a joiner in civvy street, Dibnah. Get the place fixed up, man, and keep it that way.'

This looked a good plan, said Fred: 'Wonderful, no square-bashing or early morning guard duty, just work on the tools. I was given a couple of mates to help me and we then spent most of the next year or so putting the roof back on the farm and fixing up the barns.'

When it was all done the officers moved in a dozen or so hounds, for the purpose of hunting wild boar. 'Know anything about doggies, Dibnah?' asked the CO. 'I couldn't say "yes, Sir" quick enough,' he recalled. 'Even though I hadn't come across many hunting dogs back home in Bolton, this looked like being a great skive and meal ticket, to boot. I had, up to then, never tasted wild boar.

'What made the whole Army period half-tolerable were the letters from my mam, God bless her, with news of local happenings but, more importantly, the odd postal order, the proceeds of which I squandered on luxuries like decent shaving soap and beer that didn't fight back as you swallowed it.

'But what her letters did tell me was how the steeplejacking was going on back home and who was doing what. Mind you, I didn't think Bolton would run out of chimneys while I was playing at being Master of the Black Forest Hunt. Two years' National Service did me no harm at all and I have loved dogs ever since!'

After the Army, Fred was faced with the choice of returning to joinery or trying to establish himself as a steeplejack. He chose the latter and, building on his pre-National Service achievements, again went looking for work. The motorcycle came in very handy for it allowed Fred to get around the region. Armed with binoculars, he could examine chimneys for damage to the steel bands and, with that knowledge, hopefully persuade the owners to give him the job of repairing them.

But he recalls that things were slow and, at first, he spent only three days of the week working and the rest of the time in a suit, watch chain and all, looking for more work. He did, he confided, come very close to returning to joinery and then, as happens so many times in life, fate takes a hand, in this case from an unexpected quarter.

One thing Fred does say about the Army is that it legally brought him close to things that he'd always had a fascination with – guns. This brings us nicely to the matter of a

Fred examining the mineshaft in his back garden.

Fred working on his Aveling steam road roller.

certain canon. One of the jobs Fred had hankered after was the repair of the weather vanes on the parish church in Bolton. Best suit again donned and astride his highly polished AJS, he arrived with due trepidation at the door of the good Canon Norburn, the Vicar of Bolton.

In he went to see this very important man, in his words, 'full of apprehension and a fear of rejection'. Little did he know but he had nothing to fear; in fact, as things turned out, he needed only to have a healthy regard for the 180ft height of the church tower.

The venerable gentleman took a real shine to Fred's AJS and, what's more, the vicar had parked in his drive an immaculate 1929 Humber car complete with 'railway carriage door handles' and a Weymann fabric body. It was, said Fred 'like love at first sight – him with the motor bike and me with the car'.

So off they went, chatting ten to the dozen about the way he would fix the weathercock. How it came about Fred didn't recall but the pair got round to talking about the gangster Al Capone. Maybe it was because of the style of the vicar's car?

Then, much to Fred's delight, he started talking about guns.

'Well, vicar,' said Fred, after listening intently to his enthusiastic description of his gun collection, 'I have a 9mm Luger pistol.'

Fred was an expert at making weather vanes and he often used recycled materials.

This statement, he said 'grabbed the gentleman's attention quicker than a gold sovereign on the collecting plate of a Sunday'.

The vicar then told Fred that he had a fascination for firearms and went on to confess that he had a small stock of 9mm ammunition stored in the manse, a startling revelation which came as something of a mild shock to Fred, who said: 'Well, vicar, I am afraid that my Luger is locked away in my mother's writing desk as she doesn't like me having it, let alone it being loose around the house.'

He recalls that there was no putting off the clerical gent once he knew of the classic gun's existence. 'Can you get at it, son?' the vicar asked with a strange look in his eyes. All Fred could do was to tell it the way it was, and be damned. 'Well, I do know how to pick the lock – in case of an emergency,' the young steeplejack told him.

'Could this be classed as an emergency?' the excited vicar queried.

'It could be,' Fred replied, 'as it's Friday and my mother will be out cleaning the Gas Works offices all afternoon.'

Fred recalled that they set off there and then, the vicar in the car, together with an official-looking briefcase which, later Fred learned, contained the ammo and his pistols, and Fred on the AJS. Fred's words describing what happened next can only be described as pure theatre, and, as such, priceless.

'Picture the scene: this huge black box on wheels outside our little terraced house, curtains twitching across the street and me and vicar in the front parlour. Not a speck of dust anywhere, which would be the only plus point if my mother ever got to know that Canon Norburn had been there. So there I was, picking the lock on the dresser with a practised ease, which was not unnoticed by the cleric, while he looked on approvingly. Well, we got the gun and off we went, in that brilliant car, 'up country' to Belmont, where he knew a friendly farmer.

'The next thing it's as if the Wild West has come to Bolton, tin-can shooting of the highest order. But, until you have seen a clerical gentleman with a gun in each fist like Billy the Kid, and his cassock pulled up on one side and tucked into his trouser pocket, you have never lived. He was a crackshot and a really nice guy.

'I did, of course, get the job and, as a matter of fact, I reckon it set me up locally and I have never been short of work from that day to this. That climb, as it turned out, attracted the interest of the local papers and a half page in the *Bolton Evening News* followed with pictures, and it did our business no harm at all.

'Canon Norburn finished up being transferred to Birmingham and he passed away in that neck of the woods. As for my dear mum, she never knew that he'd been in her front parlour although, on many occasions afterwards, I was to say to her, after she had finished a particularly thorough cleaning session, 'you could bring anyone here mother, even the Vicar of Bolton!'

With regard to his TV career, Fred has no qualms in 'naming the guilty party'. It was just over 30 years ago that BBC North West reporter Alistair McDonald got wind of a repair that Fred was going to effect to the clock tower at Bolton Town Hall. Fred's account of that episode in his life has been reported many times, but these are his own words describing what he always said was the start of his 'telly career'.

'Down he comes to view the job in a very natty little sports car. I'd been told that the young "telly" man was a bit of a rock climber. Good, I thought, a 200ft climb up the side of a town hall will be just up his street. Perhaps we could do the interview up on the clock

Roger in the mineshaft.

Building the pillars to support the steam winding engine platform.

face? I had a plank going from the tower and across the face, so he arrives and climbs up the inside of the tower and suddenly there he is looking out at the work site. I am standing out on the aforementioned plank 200ft up showing off the new stone pillars we had made for the job.

'Are you coming out, then?' I asked, pointing to the plank. His reply is unprintable, as they say, but we did the interview anyway and it went out at the end of the evening news bulletin.

'A few weeks later he rang to ask if he could come and see me with a TV producer. I said why not, and told him where we'd be at the time and that was that. Well, the telly folk arrived and this guy with Alistair looked like Stewart Granger, about seven foot tall, all dressed in denim, baseball cap and huge cowboy boots.

'We chatted, or rather he talked and the rest of us listened, and it was mainly in a "media" language, which I couldn't make head nor tail of. At the end of the evening I said: "What do you want me to do, give you a ring when we are going to do something dangerous or exciting?" "No," he said, "we'll ring you." Well, that was the end of my TV career, I thought.

'Now normality had almost returned to my life when Alistair McDonald, who could never be called a quitter, had a hand in things again. He had been harassing people in BBC Oxford Road, Manchester, as a consequence of which I received a call from a very nice lady called Jean Thompson. Subsequently she arrives – and with her another Stewart Granger, only this time a smaller version and much older but nevertheless the same kind of boots, long-peaked cap and nicely tailored denim outfit.

'If anything I got on with the first would-be Adonis better than I did with this one. I attempted to show him, with the help of a pencil and sheet of paper, how steeplejacks get a ladder fixed up a chimney. We had about one-and-a-half imaginary ladders up this imaginary chimney when he waves his arms. Stop! Stop! We'll have the viewers turning off in droves; it's far too boring. Quick chat with his lady colleague and there he was, gone.

Fred talks to camera just prior to setting off on his round-Britain tour.

Fred with the mineshaft, a project he would sadly not complete.

'Blown it again, I thought. But the more I pondered, the more I was sure that the first part of any programme about "our job" would need to contain the method of putting the ladders in place. Phone rings again – it was Jean from the BBC. "Can we come over on Wednesday?" They did and we had another argument about the ladders. All of a sudden, Stewart Granger MkII shows signs of capitulation. "Where are you working?" he asks. "On a big chimney at Shaw, near Oldham," I tell him. "Jean, we will call there tomorrow and have a look for ourselves," he says, and off they go again.'

Fred recalls that he had just got to the top of the 245ft chimney when the BBC pair turned up in their Mini at 8.30am.

'That was it. "Cowboy boots" instantly got the bug. They came that first time in a Mini and, as time went by, moved on to a Porsche. I'd still got my clapped-out Land-Rover. We were up and running and, to add to things, I had just won the cup at Burtonwood Vintage Show with my road roller Betsy, so that brought the steam element into the plot.'

There were in April 2004 three main projects occupying Fred's fertile mind. The first was very much normal by his standards: the continued restoration of an Aveling & Porter convertible steam tractor, maker's number 7838. The second task was something with

Ready to go, the Aveling & Porter convertible ready for the road.

which he was well-experienced: the making of another series of TV programmes for the BBC - filmed by David Hall and his team from *The View from the North* in Leeds. However the third one, even for Fred Dibnah, was a little unusual: the design and sinking of a five-foot diameter 70ft deep mine shaft leading to a 5ft 6in-bore tunnel, the associated head gear and engine house for which he had long since had permission to build. Unusual? Well, you must suppose so when you are told it is in his back yard and possibly in a place where lesser mortals would have preferred a barbecue. The strange thing is that, when you got to know Fred, the execution of this task, which had most of the Bolton city fathers excited, seemed like a perfectly natural thing to do.

This frenzied pack of civic officials and councillors called on just about every authority they could think of to have the project permanently stopped. But all they succeeded in doing up to that moment was having a halt called to work while a review took place. It may be that, in Bolton, an Englishman's home is still his castle but it cannot, by their reasoning, be his simulated coalmine.

Fred, according to much local opinion, had a point when he claimed that, to some of the council's officials, any reminder of the area's past 'cloth cap' image was an anathema and could not be tolerated. Oh yes, they were pleased as punch that Fred's work had been recognised by the awarding of an MBE.

'Isn't that good for the local image?' you can imagine them saying. But can we attempt to remind the up-and-coming generation of Boltonians of their heritage in those far-off days when little children were sent down into the ground to dig coal? Dear me, no, that would never do.

Fred surmised at the time: 'It could be that they are afraid of me becoming a tourist attraction. If so, that's rubbish. Already parties of youngsters have attended our workshop to learn something about the ways of the past. There is hardly going to be more of them; it's just another bit of history for them to be reminded of.'

Thankfully millions of people are now aware of the saga following the penultimate TV series; it has even (already) been repeated because of the interest it generated.

Who, having seen it, will forget the bow and arrow incident. It may have looked pure fantasy but it was based on sound engineering practice.

The skill of creating a working size bore using an iron 'sinking ring' was fully explained in the series. That age-old technique was something many people had either never heard of or else had forgotten about. A good team could, by Fred's reckoning, advance the shaft's depth by as much as two feet in a working day.

The plan called for the shaft to join with the slightly inclined tunnel and could, with the addition of railway lines and the utilisation of four trucks (coal tubs coupled in two pairs), be able to demonstrate the way in which coal was raised to the surface. The work had progressed steadily over several months, as time permitted between steeplejacking, filming for TV, public appearances and working on the 'convertible', and then came the order to stop work.

Some well-meaning soul, who must have stood on tip-toes on the top of Fred's adequate boundary fence to even see the work, reported the matter to the council. The publicity created a growing pile of correspondence from local people in support of the venture, some of them very eminent persons and, as Fred was taken ill, the so-called Darcy Lever Mining Company was preparing to go to appeal to have the work cessation order lifted.

One of the parties called to inspect Fred's pit shaft were the Mines Safety & Rescue people from Selby in Yorkshire. They found nothing wrong; on the contrary, they thought

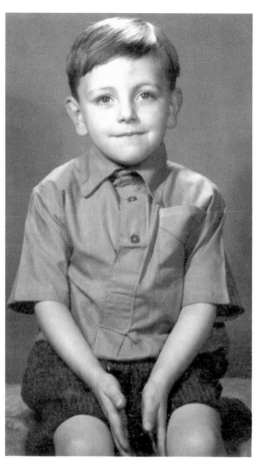

Fred the schoolboy.

Dr Fred Dibnah. Fred after receiving his honorary doctorate from the University of Birmingham.

Fred the soldier.

Fred on the top of Bolton Parish Church.

it was a very good idea and looked forward to seeing it finished. The council also called in geological experts and that inspection had an amusing outcome. Their claim inferred that the hillside on which Fred had started sinking the shaft would become unstable because of the work. In fact, they reported that, since Fred had started sinking the shaft and brick-lining it, he had made the hillside much more stable than it previously was!

The Aveling & Porter convertible compound steam tractor, which Fred had been in the process of restoring, or perhaps more accurately, totally rebuilding, was a real labour of love that had been ongoing for the last twenty-seven years. In the spring of 2004 it was, in Mr Dibnah's words, 'just about ready to embark on a grand tour of the UK'. It was to be coupled to a fine, newly reconstructed living van that was rescued as severely fire-damaged from a field in Burnley. The superbly finished engine would go on to share top billing in the new TV series and, as a part of this, it would be proudly driven around the streets of the nation's capital.

The compound was originally built for Somerset County Council in 1912 and configured as a tractor. However, by 1914 it had been commandeered by the War Department and sent off to France in the form of a road roller. It came back war surplus and was used for all of its working life as a roller by Devon County Council. It was, towards the end of that life, parked in a layby where someone stole all the brass fittings from it. It was then advertised for sale by tender.

It was subsequently bought, those forty years ago, by a friend of Fred's for only £200. He said that, at the time, he was green with envy, having just paid £175 for his roller. The machine (registered TA 2436), looked to be in very good order but, on closer inspection, that was found not to be entirely the case.

Several years later, in 1977, Fred bought the convertible and moved it to his Bolton workshop. At that time he said it looked very good, lagging on the boiler etc, belly tanks all lined out and a smart paint job. But there were tell-tale signs for all to see including a big L-shaped weld on the side of the boiler right underneath the point where the high-pressure valve spindle is.

The standard of the riveting on the boiler also left a lot to be desired. It quickly became obvious to Fred that this boiler barrel would never pass muster and so the twenty-seven-year slog began. The front axle was found not to be original and was, in fact, from a Garret No 4 tractor, while the front wheels were also the wrong ones and were even of different diameters. They obviously needed to be replaced by a suitable Aveling pair, and new bearings fitted, before the restoration could be deemed complete.

The tender, belly tanks, boiler barrel and ash pan were all replaced by newly manufactured substitute parts but the biggest job by far was the fabrication of a new throatplate. X-ray examination had revealed no less than twenty-seven cracks in the old one. In helping with the manufacture of a new one, Fred picked out the Hyde-based firm of Bowns for special praise. They are, he said at the time, 'the kind of outfit that steam men should support'.

Having been privileged to study up close this remarkable Lancastrian and witness first-hand his grasp of so many diverse engineering techniques, there is only one conclusion to draw: Fred Dibnah would have made a success of whatever business he'd chosen. Come to think of it, what should he have put on his passport – joiner, steeplejack, engineer, mechanic (back street or otherwise), storyteller, mining contractor, television presenter or just plain Great Briton?

Following Fred's untimely death, one of his admirers wrote in answer to that question: 'None of the above; try national treasure.' I am sure we would all like to be associated with the remarks of that gentleman.

During a visit to the Great Central Preserved Railway, not long after receiving his MBE, Fred suddenly produced the medal from his overall pocket declaring to the delighted onlookers, 'I suppose this means I outrank you lot!'

A Capital Affair
Fred Collects his MBE

To stand on the corner of Birdcage Walk and Parliament Square in the rush hour and witness the passing of an Aveling & Porter convertible steam tractor and living van is one thing; to hear the cry: 'Rook, it's Fled Dribner' exclaimed in unison by an excited group of tourists from the Orient is entirely something else.

Both these happenings will not be quickly forgotten by those fortunate enough to have been on hand to witness them.

To cap it all, diametrically-opposed technological happenings became almost common-place as the day unfolded. More specifically, it was the capture of images of a 1914 steam engine on the microchips of state-of-the-art 'picture' phones, their excited owners despatching those once-in-a-lifetime pictures to contacts worldwide.

It would, we had mused on the journey south, be interesting to know if the populace of this overcrowded and manic city would be familiar with the highly polished art of our Fred. There was no better place to start than with a member of that special breed of capital-dwellers, the London cabbie, our pilot on a journey from Euston to the appointed meeting-place. Not quite the perfect answer, but not bad for a start: 'Lov 'im, straight I do, gov!

The Irish Guards were host to the engine and film crew at Wellington Barracks in Birdcage Walk. Fred thanks the officer in charge.

Roger has a practice drive at Wellington Barracks.

Especially the bit abart the mine in his street. They live a different life up there in Yorkshire, straight they do.'

We paid the fare and tactfully corrected his geography, but it was not bad for a start.

Our next 'pilot of the bus lanes' was a gentleman definitely not born within the sound of Bow bells. In fact he told us, though unnecessarily on account of his rich nasal accent, that he was a Scouser. Apart from trying to claim Fred's birthright for Maggie May's city, he declared that he was an avid watcher of the TV programmes and even confessed to having his lady tape them when he was on late shift.

A Transport For London bus inspector was next. 'Leave off, mate! Who ain't seen the trick wiv the bow and arra. What was that all abart?' We put him wise and made a sharp exit. During a break in filming, the crew swapped stories and it seemed that they had all had similar experiences when asking about our boy, save for one gentleman who thought Fred was in a boy band. Come to think of it, if called to he could well play that role successfully!

So now the only worry was whether 'Her indoors up the big house' would recognise him when the time came. After all, the keeper of her parks had forbidden a perambulation up The Mall with the little compound. The plan for Fred to arrive at 'Buck House' via The Mall on the engine was unfortunately scuppered by officials who declared that the weight of the engine and living van would be too great for their road system.

Returning to the barracks after the parade around London.

Before and after the formalities, a celebratory run around the capital was undertaken with a rush-hour trip up Birdcage Walk, direction Parliament Square. The tourists get a sight of the Aveling & Porter in the morning sunshine and the TVFTN film crew are in the living van.

Whitehall.

The clothes, they do say, maketh the man. Either way, it's still our Fred, as he emerges from the living van to applause all round from the crew and the watching soldiers on the parade ground at Wellington Barracks.

However, help was at hand and the Adjutant to the Irish Guards, Alex Turner, was thankfully able to offer the crew the facilities of Wellington Barracks, a location within easy walking distance of the palace. Even better was the fact that the Aveling and Alan Atkinson's low-loader would, as a consequence, have a really secure resting-place while in town.

The Great British icon – and a Great Briton.

Jack and Alan have a well-earned breakfast sandwich after lighting up the engine prior to the run around the capital.

The route chosen for the 'victory parade' (and that term is justified if the constant cheers, waves and hooting of car horns was anything to go by) started with a trip up Birdcage Walk followed by a circuit of Parliament Square and Westminster Green. Thereafter the Houses of Parliament were circled in an anti-clockwise manner via Millbank and then a crossing of Lambeth Bridge being followed by a smart run up Lambeth Palace Road and a left on and over Westminster Bridge.

The next circuit, necessitating a sharp right turn off the bridge and in sight of the London Eye, saw the 'Bolton Wanderers' travel in a northerly direction along Victoria Embankment. A left just before the Hungerford railway bridge brought our travellers into Northumberland Avenue and then, by hanging a left at Nelson's Column, to Whitehall and up past the Cenotaph. Whitehall brought the boys back to Parliament Square, at which point the watching public were treated to the sight of the Aveling & Porter making a further circumnavigation of the whole route including another 'pass' of the Palace of Westminster.

By this time those oriental tourists were quite ecstatic and seemed to have doubled in number. A brief glimpse of 'Red Ken' on the Green as he talked to a party of placard-waving demonstrators seemed to be cut short by a sharp shower of rain. Some wag on our crew asked if that nice Mr Livingstone had really come to see if Fred had paid the London Congestion Charge. And, if you're wondering, yes, he had. Or at least Alan had paid it for him.

Fred and the Major.

Northumberland Avenue, out of all the roads on the route, had seemed to stick in Fred's memory and the reason was to become obvious. It was approaching lunchtime when a voice on the 'ship-to-shore' communication system, to wit Big Alf and David Hall the producer's mobile phone link, announced that a very nice hostelry with great windows and a black and white timbered façade had been spotted on a previous circuit.

This fine watering-hole turned out to be the Sherlock Holmes public house at the northern end of Northumberland Avenue. 'Can we park up OK?' was the logical question from the crew chief. 'Watch me,' was the affirmative reply from 'the man' on the lean, mean, green steam machine.

Now coincidences, some would have you believe, are more common than enough while others would swear that the opposite is true. As all settled down to a quiet sojourn at the pavement tables outside the pub (our summer had abated and sunshine had fleetingly returned), one such occurrence was to take place.

A very strong military-type voice with a 'clipped' delivery momentarily broke the tranquillity of the scene: 'You have got no taller, Mr Dibnah!'

'And you no smaller, Major,' replied Fred as quick as a flash. A watching member of the crew was quickly despatched inside the bar to procure a vessel for the officer and he joined the lunch party. Now here is the coincidence bit. Given that London is a fair big town, even by Bolton's standard, what was the chance of Fred meeting up again with the man who, in his words, 'had queued behind him at the Palace'?

Major Richard Courteney-Harris, a serving officer with the Queen's Lancashire Regiment, had received an MBE at the same ceremony as Fred. He was out walking the town with his delightful wife, Karen, and their children when he just happened upon the Sherlock Holmes. A nice reunion? Well, of course. Elementary, my dear reader.

While this impromptu MBE celebration party was in full cry, the attention of the group was drawn to the parked-up engine and van. 'Something is occurring,' said Big Alf and the situation could, he said, 'require the tact, diplomacy and charm of Fred'. Now traffic wardens anywhere are a breed unto themselves, but the London contingent of that calling are something special.

A lot of head-scratching and muttering was taking place on the footpath at which the steamer was parked. These actions were followed by a guy with a shuffling walk circling around the ensemble while casting steely glares at the double yellow lines. This activity culminated in a question or two. 'What is it and what's it doing parked here?' Remarks such as 'bet his clamp won't fit that' and 'throw your booking pad in the fire, mate' were emanating from onlookers who had assembled.

Fred arrived on the scene. He told the custodian of the yellow lines that 'the driver of a steam conveyance is entitled to park anywhere in order to attend to his fire or take water'. He further explained that a failure to do so would result in a huge bang as a result of which the local glazier might need to pay a visit to Northumberland Avenue. The dear man was charmed and hastened down the street declaring 'leave it wiv ya, mate!'

Accordingly it could be said that the Dibnahs and their steam tractor had charmed the capital. Fred thoroughly enjoyed his well-deserved visit to the palace and *The View From the North* got the whole thing on film. Summing up that great day, the only thing that remained to be said, pinching a quote from Julius Caesar, was: Veni, Vidi, Vici'.

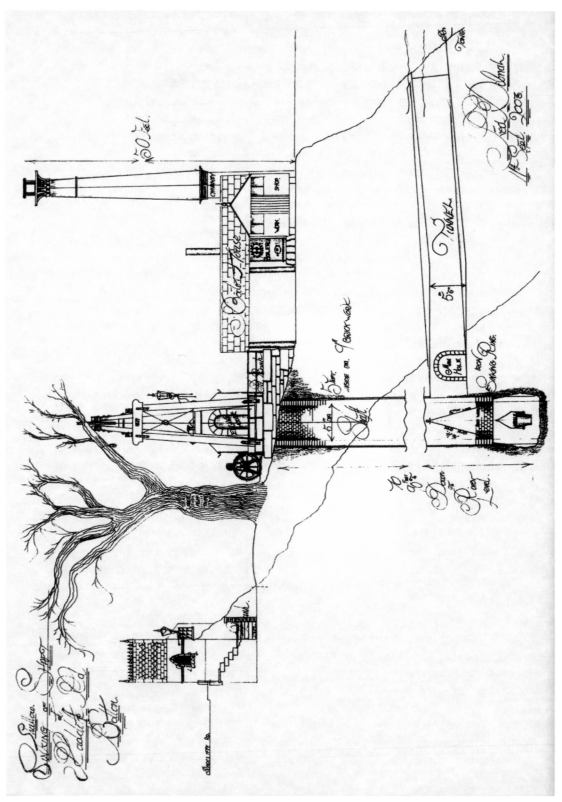

Sinking of Mine Shaft at Radcliffe Road, Bolton, Fred Dibnah 11 September 2003.

Fred gets to work with the polishing rag before travelling down the Llanberis Pass.

Eh! Mister any chance of a go at driving this engine?

Alf Molyneux with Fred.

Bolton School of Art

During the incident of the shocking discovery in the Bolton canal, Fred was a pupil at art school. A classmate at that time – 1951-53 – Edward Williamson, takes up the story, saying: 'Fred was forever going off exploring, as he put it, and so when the buzz went around the school that the police had come and Fred Dibnah had been taken off to the head's study, we were all agog with excitement and started to imagine all sorts of things.

'He was missing from the class for a couple of hours.

'At that time we knew nothing about the body in the canal as Fred had been told by the police to keep quiet, and he'd done just that.

'Rumours started to circulate freely at break and a good many of the pupils had Fred bang to rights on the charge of committing a very heinous, but of course imaginary, crime, but he soon joined us and was quick to tell us all about the event.

'He was full of it, gruesome details and all. He told the tale in a very graphic manner, similar to that which was to make him such a great storyteller and broadcaster in his later life.'

Edward said Fred's love of exploration was such that, if you took off his trademark flat cap and put a battered trilby hat on his head, you would have invented a real life Indiana Jones, Bolton-style!

He said their teachers would ask Fred for suggestions about the best locations for the class to visit for drawing practice and Fred always knew which canal bridge or chimney

Fred's art school friend, retired joiner Edward Williamson, now spends much of his time making models. He is pictured here in his workroom.

would make a good subject. The teachers were never disappointed with his locations.

Fred, Edward said, was so well informed about the local area because he had explored just about every square foot of it with his regular circle of mates at some time or other.

'We knew what the road over a particular bridge looked like. The thing with Fred was that he also knew what it looked like underneath, because he had inevitably been to have a look!' He added that Fred did take them to some great places and recalled that the old canal figured strongly in many of those ramblings.

Edward and Fred had two separate stints together at the art school. They both first did a two-year art course and then left to start work. However, both having chosen joinery, they were reunited for a further spell at the college as part of their apprenticeships, to learn more drawing techniques and the theory of woodworking. Edward has saved some old sketches by Fred and has allowed us to reproduce sections of them in this book.

Edward recalls that he had a big waterproof satchel at the time while Fred did not, so he often carried home some of Fred's work along with his own in order to keep it dry.

From reading the teacher's comments, it is easy to see that Fred's flair for elaborate design was evident even at that early age. Such comments as 'very nice but a little too

A thoughtful Fred.

Bolton School of Art in 1953. Young Fred is seen second from the left on the second row from the front. Who could mistake that cheeky grin? Edward Williamson is on the far right of the third row (behind the youngster in the white shirt), while Donald Jackson is on the extreme left of the third row above Fred's shoulder. Lonsdale Bonner, the teacher who helped Fred set up his steeplejack business, is seated third from the left on the front row.

intricate for a mere bathroom stool', were made about one project which gained a mark of 16/20, while a 'good drawing but not what you were asked for' was given 17/25.

Another classmate was Donald Jackson from Leigh who, after art school, found employment as a calligrapher and went on to achieve great things. He became the official scribe to the Crown Office at the House of Lords and his work became well-known not only in this country but also in the USA – not bad for a Lancashire lad who used to share Fred's interests in shooting and bird-nesting!

There was another twist of fate which brought the two Bolton men's lives together on another occasion. They both went off to do National Service in the Army but served in vastly different parts of the world, Fred in Germany and Edward in north Africa.

Edward served as a joiner during his posting in the company of another Bolton man, Ron Crossley. A week or so after they were demobilised, he met up with his Army pal in Bolton and, during the course of their reunion, Ron told him that he had got some work. When Edward asked him what it was, he was told: 'With a steeplejack called Fred Dibnah.' Edward spent his working life in the building trade. He is still, so to speak, chopping up wood in his retirement. His old pal Fred would be proud of him, as he is a very accomplished model maker. Come to think of it, the pair's teacher at the Bolton School of Art wouldn't be too disappointed either! That educational establishment no longer exists as such but the building that housed it survives, albeit as flats.

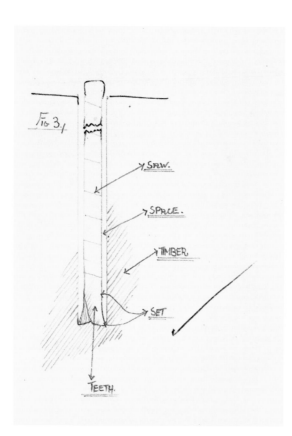

Fig 3.

SAW.

SPACE.

TIMBER

SET

TEETH.

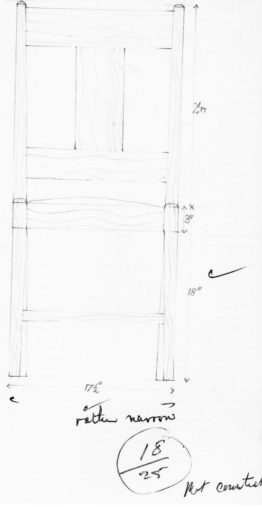

2 FT

x x
3"

18"

17½"

rather narrow

18
25

Not counted

Fred had two spells at the Bolton School of Art and these drawings were done by him during his second spell as a woodwork student. Note some of the teacher's comments; it appeared that Fred liked embellishing his work – something it appears was not always appreciated by the schoolmaster.

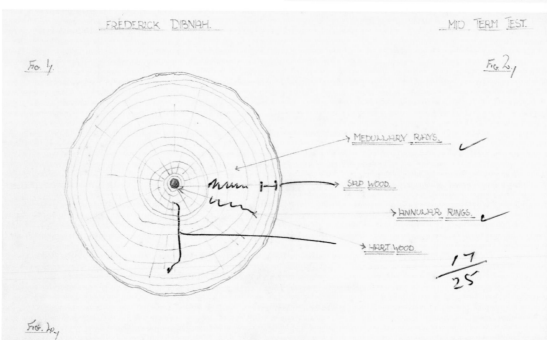

FREDERICK DIBNAH.

MID TERM TEST.

Fig 4.

Fig 2.

MEDULLARY RAYS.

SAP WOOD.

ANNULAR RINGS.

HART WOOD.

17
25

Fig 2.

'Betsy is an Aveling & Porter Ltd 'BHO Class' single cylinder 5 nominal horse power 10 ton road roller, dating from February 1912 and carrying works number 7632 and road number DM3079. Fred was often to be seen out and about with the roller, which he named after his mother.

The Finishers Arms, seemed as good a place as any to stop for water!

'Betsy' was a regular attendee at steam and vintage shows far and wide. Fred and his young family are seen with the Ex Flintshire County Council roller at such an event.

...Fred is seen deep in concentration during the reconstruction of his beloved road roller.

...Fred on board, prior to a road run.

...A 1960's view of Fred's roller being driven passed an admiring group of young Boltonians.

Summer 2004

During the times Fred invited me to visit, or travel with him, I was able to take literally hundreds of pictures. They were originally to become the basis of a series of magazine articles about the man, and the close contact he enjoyed with his adoring public. Choosing images for publication is a difficult enough job, but with Fred more so, because you just wanted to use them all, and of course there was never enough space. What follows is a selection of the 'other pictures' all taken during Fred's last months in the summer of 2004.

You missed a bit Alf!

As Fred is studying some old pictures of Anderton Boat Lift, prior to filming there, he is besieged by a party of lady day trippers.

Fred at Anderton Boat Lift

Ready to start filming, next to the Trent & Mersey Canal.

Filming at Anderton, Fred always maintained that the boat lift was one of the greatest feats of engineering he had ever seen.

The talk is about waterways engineering and Fred listens intently.

Alan Atkinson's mobile Low Loader café is now closed, breakfast is over and the boys must get to work polishing the engine. Note three Dibnahs but only two chairs!

Fred at Anderton Boat Lift.

The Dibnah boys chat with Gary Hughes of British Waterway.

The leafy lanes of Cheshire make a fine back drop for Fred's Aveling & Porter convertible traction engine. The engine was built in 1912 (just like his roller) and it carries the makers number 7838 and the road number TA2436.

Fred and Alf are on a visit to a local foundry.

Fred eases the 'convertible' and living van through his gate and onto the road for the first time.

Tripping around Bolton.

Proud as punch - first trip with the Aveling

Fred dropped his last chimney on Sunday, 9 May 2004 at Lion Mill Royton near Oldham. The young admirer has just collected Fred's autograph.

A lady autograph hunter gets her man.

Fred and Jack discuss a modification which they needed to make to the firebox door of the 'convertible'.

Resting between 'takes' at the Great Central Railway, left to right Roger, Fred, Alf Molyneux and Jimmy Crooks.

Learning the script at the GCR.

The firebox door on the Aveling tractor continues to give trouble. Fred and the boys discuss another way to solve the problem.

A last 'mug up' before the cameras roll.

The firebox door job is done, Jack waits to inform his Dad that all is well.

Fred gets to ride the steam locomotive footplate at the GCR.

Fred and Phil Jeffs are joined by Jack and Jimmy Crooks.

The steam driven saw cuts the brake blocks at the Klondyke.

b done, time for a pint

Fred as he would have liked us all to remember him.

uring the Klondyke visit a tree was felled, and from the mber a new set of brake blocks for the convertible were ashioned. Here Fred is writing down the necessary imensions on a sliver of tree bark.

During a visit to Klondyke Mill, Draycot in the Clay, Staffordshire young Jack tries unsuccessfully to relieve his Dad of a pint of the 'Black Stuff'.

Driving practice for the boys. No better place than the parade ground at Kensington Barracks, Birdcage Walk, London.

Father and sons outside the Sherlock Holmes public house.

Filming in London. Fred talks to the camera in his 'palace suit' Picture taken from the 'wings'.

Houses of Parliament again, and its still raining.

Parked in Whitehall. 'Veni Vidi Veci' said Fred, or something like that anyway!

The tractor and van are steered carefully into Kensington Barracks, after filming 'down town'.

Fred looks at a locomotive boiler tube plate under repair, he was impressed and said so.

A visit to the Midland Railway Centre

The smiling 'Three Amigos' return, what they have spotted was mine headgear. Alf and Jimmy are both Ex coal miners.

Alf realises it's his plastic which will be used to buy the lunches, Fred acts coy!

Fred enjoyed his visit to the steam locomotive repair shops at the MRC.

A visit to the Severn Valley Railway

On arrival at the Severn Valley Railway Fred and Alf made straight for the tea shop, much to the delight of the other customers.

A quiet moment at the end of the day with a small libation, good man Fred!

Fred takes a moment on the foot bridge at Bridgnorth Station SVR to enjoy a view of the engine sheds and savour the smell of smoke, steam and hot oil, sheer magic.

Not a lot of parking spaces at the top of the Llanberis Pass! Accordingly Alan Atkinson drops Fred off at the bus stop, before trying to find a place to unload the engine.

Tripping around Low Town, Bridgnorth.

A kettle is a kettle! Fred chats with the steam locomotive engineers at the Severn Valley Railway.

'That's some hole in the ground' Fred gets his first view of the once active opencast copper workings at Parys Mountain, Anglesey north Wales.

Fred and Alf spot a certain photographer covered in mud. I had fallen into the ditch whilst setting myself for this shot!

Roger said he could walk up the hill quicker than Fred and the tractor. As they reach the top Fred eases back so not to disappoint the youngster, then he opened her up and blasted to the top arriving first!

Fred and his new living van– first time on the road

Fred tows the living van up to the road for the first time prior to its run to Astley Green Colliery.

Alf and Jimmy steer from the back as the van finally reaches the road, Fred on the Aveling.

In Bolton town centre for the first time with the van in tow, the boys head off at the start of the 'Made in Britain' tour.

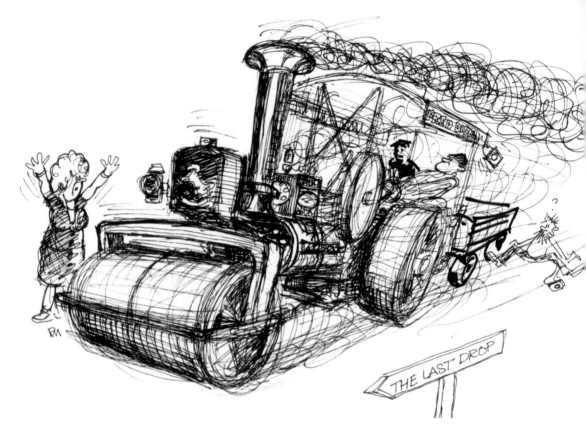

The Last Drop Disaster

Retired architect Bill Greenhalgh was associated with Fred for a length of time he describes as being 'more years than enough' and, like all of Fred's associates, has many a tale to tell. He also admits that between them they 'probably stopped a few barrels of ale going bad', a fact which, if true, would account for his unique claim. He had the distinction, he says, of being banned from the Dibnah homestead by not just one but, in turn, all three of Fred's wives!

He claims he was always the fall guy. Whenever Fred was late arriving back indoors, or was otherwise uncontactable and seemingly adrift, his name would be proffered by way of an excuse. These disqualifications led Bill on one occasion, during a particularly long banishment, to resort to drastic measures in order to contact his pal. This was long before the proliferation of mobile phones and Fred probably wouldn't have had any truck with one of those anyway.

Desperate to get in touch, he'd positioned himself across the other side of the River Tong at the rear of Fred's workshops in an attempt to make surreptitious contact, but all to no avail. Try as he might, he could not get Fred to look in his direction, and then he hit upon an idea. He hurried home and, retrieving some scrap timber from his shed, he made a placard board.

He fixed a long broom handle to the board, on which he had painted the words 'Hello Fred, How Are You' in letters six inches high. He then returned, this time to the front of the Dibnah house and paraded back and forth on Radcliffe Road, with his placard held high above the perimeter fence. This ruse did the trick: he was spotted and temporarily forgiven

Bill Greenhalgh by the rebuilt brick pillar as he recalls that fateful day in 1974.

The roller started sliding at the point where the Demdyke Suite sign is now situated.

for leading Fred astray. He recalls that, after laughs all round, 'normal service was quickly resumed'.

The Last Drop disaster

Though others know of, and are associated with, the happenings on this quiet October Sunday in 1974, Bill wanted to be the first to recount the tale. There is good reason for that as, for more than thirty years, he has had to get used to being taunted as the man who bailed out. But that is, of course, part of the end of the story; the beginning started, naturally enough, with a chimney.

The Last Drop village is situated on the edge of the Pennine Moors above Bolton and consists of a collection of eighteenth century farmhouses looking for all the world like a hamlet lost in time. But The Last Drop is home to one of the most prestigious hotels in the north-west of England and is owned and operated by Macdonald Hotels & Resorts. In addition to the hotel proper, there are cottages, gardens, courtyards, craft shops and a delightful village tea shop. In these modern times the village boasts extensive leisure and beauty facilities and it is also a popular venue for weddings and functions.

Back in the 1970s the owners engaged Bill Greenhalgh as a building designer, as a consequence of which he was heavily involved in the development of this complex. Rebuilding one section of the old farm, in what was to become the then North West Tourist Board's HQ but is now guest bedrooms, Bill's plans called for materials sympathetic to the surrounding buildings to be used. In line with that policy, he was looking for new chimneystacks – new, that is, in relation to that site, but old style was, of course, his preferred option.

Who else would Bill turn to in order to locate the necessary materials than his long-standing friend. Fred Dibnah was, it transpired, at that time dealing with the demolition of

The chimneystack, which is effectively to blame for the disaster.

a Victorian mill-owner's house on the other side of town. Just maybe, thought Bill, that site could produce the goods. With enthusiasm and relishing the excuse to team up again with his steeplejack friend, he set off in search of the demolition site.

It was a very fruitful journey and Fred was easily persuaded to remove very carefully a couple of Victorian chimneystacks, after a suitable fee had been agreed. The stacks were made up of worked stone slabs octagonal in shape and about two feet across, each section being eighteen inches thick. When assembled in the right order, the stones formed the shape of the chimney, and each had a nine-inch flue hole in the centre. Given that, when erected, the structures were over six feet tall, the resultant pile of reclaimed Victoriana weighed over twenty tonnes.

Fred had been using his steam roller during the demolition job to pull down sections of masonry, so it was a natural conversational progression that brought the two men to agree to use the roller, and a suitable trailer, to transport the stone sections across town and up the steep hill to the village. The stone slabs had been loaded and secured on the trailer at the end of work on the previous day. Down at town level, on what started as a bright and sunny Sunday morning, Fred needed only to steam up the Aveling, hitch up the load and set off on the climb to the village.

The outward journey was uneventful, with the steam roller performing magnificently. Therefore, even at a speed of only 4mph, Fred and his crew of Bill Greenhalgh and Michael Webber reached the hotel site in good time and with the load intact, having made just one stop for water. However, on the way up they were aware that the nice early morning weather was rapidly being replaced by damper conditions, a change made more apparent by the lack of boards on the canopy roof of the roller. The uprights were in place and the side boards fixed but the roof planks were still lying in Fred's yard.

There was, in those days, a pub on site and, surprise, surprise, it was called The Drop Inn. The first priority, however, was to carefully offload the stone slabs as near as possible to the building of which they were intended to form part. The unloading was hard and therefore thirsty work, so it was not surprising that, upon its completion, the boys retired to the bar of The Drop Inn.

They do say in that area of Lancashire that, if you can't see Winter Hill clearly, it's probably going to rain, and if you can't see it at all, then it's appen' already raining. Consequently, the Pennine weather has a habit of getting worse before it gets better, which is precisely what happened while the trio were taking their well-earned refreshment. In fact, Bill recalls, the light wind had turned into a moderate one and was shaking the leaves from the sodden trees. as they left the bar.

For the journey back to Bolton, Michael joined Fred on the footplate as it was his turn to steer. Bill, being relegated to hang on the back, instead opted to ride on the trailer, probably one of the best decisions he has ever made. Owing to the change in the weather and the lack of a canopy, the three were soon soaked to the skin. But otherwise, with a good fire in the roller, plenty of water in the tanks and the steam pressure adequately lifting the needle, all looked well for the downhill homeward journey.

The ensemble first travelled along the central road of the complex before making a sharp right-angled turn to the lane leading out of the village. The boys were happy and generally chatting about a job well done.

Following what happened next, Bill has always said that he has sympathy with Britain's much-maligned railway operators.

During the first twenty yards of their 300-yard journey to the main road, on a falling grade of about 1 in 80, it soon became apparent that all was not well. The newly-metalled lane was coated with wet, damp leaves and, accordingly, its surface resembled more an ice rink than a road.

All of a sudden and without warning it happened. The Aveling and trailer started to slide, there was no holding her, and the steam roller developed a mind of her own. She got into a slide that Bill says "would have done credit to Torville and Dean" and headed down the incline, gathering speed at an alarming rate.

The steamer and trailer were by this time completely out of control and, at a point about fifty yards into the descent, Bill decided that enough was enough and he jumped, diving over a stone wall and into a field, paratrooper-style. Fred frantically threw the engine into reverse and Michael manfully juggled the steering, but all to no avail. In reverse the driven rear wheels seemed only to hasten their descent and add to the slipping effect.

Having recovered from the shock and jumping quickly to his feet, Bill observed a spectacle he describes as being like something out of the great chariot race in *Ben Hur* occurring before his eyes on the damp Pennine hillside. It seemed, he said, to be happening in slow motion, with Fred and Michael hanging on for dear life in poses reminiscent of those ancient charioteers.

To the left was a very substantial stone wall, to the right and down a steep bank was the perimeter wall of a hospital and in the centre was a huge stone pillar. Whether by accident or design the roller, after first clattering over a cattle grid and demolishing a five-barred gate, then struck the solid stone pillar dead centre.

In a cloud of smoke and steam and with a sound like crashing thunder, the machine finished up elevated at 45 degrees with its front end perched on top of the rapidly crumbling mound. The actual roller was thrown to one side after the impact, on account of the forks having been smashed. The two charioteers had clung on to the end and, although very shaken, they were thankfully unhurt.

Bill scrambled back over the wall and ran towards the scene of the crash only to see the roller take another lunge forward, which that time resulted in it settling into an elevation of 45 degrees the other way around. Fred and Michael were frantically trying to 'throw out' the fire without burning themselves – not an easy task while trying to keep their balance on the precariously angled, crashed machine at the same time. On top of everything else, what they didn't want was a blown boiler.

Calmness returned to the scene and Bill recalls that, with the carnage now becoming apparent and the roller looking very much the loser in the battle with the brick pillar, the owner of the hotel (his then boss) came up the drive in his car. As the three stood rooted to the spot, he simply opened his car window, nodded at the wreck and after remarking 'silly buggers', drove off. At just about that time a resident from one of the nearby cottages, who had apparently been having his Sunday tea in his front room and had observed the spectacle first-hand, appeared with a camera.

'Yeh! Yeh!' said Fred enthusiastically, in answer to his request to record the scene, 'take as many pictures as you like but don't send 'em to the bloody Bolton Evening News, cock.'

The boys, having made the wreck safe and cleared the debris from the roadway, covered the poor steam roller with a borrowed tarpaulin sheet and went home to contemplate their next move.

Always one to see an opportunity in every problem, Fred recalled that the forks were not the right ones for the roller anyway so this mishap would be as good a time as any to change them. He recalled that a guy in Yorkshire had once offered him some, so there ensued a journey over the hills that resulted in the acquisition of a new pair of 'proper' forks.

Lifting gear was transported up to the crash site, allowing the Aveling to be repaired where it sat, still at the bottom of Last Drop Lane, a week or so later. Remarkably, it was then steamed and driven back home. But, as you may guess, that was not quite the end of the story.

As the boys prepared to set off back down the hill, a young lad came up to them waving half of the 'Invicta' plate which had broke off during the impact. 'Can I have the other half, mister?' he said, pointing to the bent half-sign still screwed to the roller.

'Yes, son, you can,' said Fred whipping a screwdriver out of his pocket, 'but only if you tell me who was the guy who took the pictures.'

'He lives in the bungalow over there, the one with the caravan in the drive,' he said as he skipped happily away clutching the two halves of the Invicta horse.

Fred knocked on the door to enquire about the pictures. The 'Law of Sod' then entered the equation. Before the days of TTL (through the lens) viewfinders on cameras and when using 'manual wind-on', you could shoot away happily with the lens cap still 'on', which is just what the guy had done. No pictures.

The only record that remains of the incident is the one in the minds of the two guys who were there at the time, but the chimneys which started it all, are still there to be clearly seen. The Last Drop village is a great place to visit. If you do, it is important to remember that the lane concerned is the one to the extreme right of the hotel as you view it from the main road. It now has 'speed bumps' – not that they would have been any deterrent then, mind! About where the second bump is now, is where Bill Greenhalgh jumped.

Were you that young boy who won the two halves of Fred's Invicta plate and, if so, do you still have them?

Fred with mining historian and friend Alan Davies, deep in conversation discussing work in progress on the restoration of the Aveling tractor.

The Eyes Have It

Following on from the tale of Fred, the BBC and Bolton Town Hall, there is another tale concerning derring do and matters of a civic nature that needs telling, although a delicate touch may be the order of the day.

The somewhat unusual anecdote was recounted for all our benefit by Bill Greenhalgh, with whom Mr Dibnah got into what have been described as 'a few scrapes', which I suspect is one of the biggest understatements of all time!

But back to Bolton's major civic edifice. The original town hall, of which Fred was fiercely proud, was opened on 5 June 1873 by HRH Albert, Prince of Wales. There was a hall in that fine building named in honour of the young prince but sadly, the original majestic building was ravaged by fire on 14 November 1981.

The conflagration reportedly started in the basement and destroyed the original Albert Hall but, as fortune would have it, the rest of the building was saved. The damaged section was later rebuilt as the new Albert Hall and Festival Hall, both of which are used for functions.

There is more than a little irony where the creation of Bolton's town hall and the message in the town's motto is concerned.

The civic motto is 'Supera Moras' which literally translates as 'Overcome delays' but the time that lapsed between the original town hall proposal being submitted and the completion of the building was the thick end of seventy-seven years! However, as someone's granny in those parts is sure to have said at some time or other: 'There's nowt got with rushing – only babies!'

Bill Greenhalgh is known locally as a designer of buildings but it was also rumoured that he did get his hands dirty on occasions, and this is an account of one of those occurrences.

Fred carried out a fair bit of work at the town hall during the late 1970s including famously the gilding of the globe, which first brought him to the attention of the BBC. He always said that the civic jobs were not big payers, more labours of love. One such labour was repairing the columns to the cupola (a small dome above the central roof).

Fred having a quick fag while giving the job a coat of looking over.

Bill Greenhalgh's superb watercolour of Bolton town centre, showing the town hall with its cupola.

The council, no doubt with one hand firmly on the purse strings, had asked Fred to repair/replace the pillars using glass fibre, a request made during a site meeting at altitude.

At that suggestion, Fred reportedly went ape and marked the council engineer's card in good style. He left the poor guy in no doubt that only 'proper' stone would do and that he – Fred Dibnah – would not put his name to what he said would be 'a right shoddy job using that new fangled fibre stuff!'

There was no more to be said. Doubtless the council guy saw discretion as the better part of valour and diplomatically climbed down to terra firma – but only after telling Fred to get the job done as he saw fit.

Having recruited Bill Greenhalgh to help him, Fred completely bamboozled his friend by first sending him to the scrapyard in search of an old washing machine.

Why, thought Bill, as he haggled with the scrapper over a couple of bob, did Fred want a bloody washing machine? On arrival back at the yard, Fred asked Bill if the one he had procured had a good motor?

Funnily enough the scrapyard man had said that it did, but that the pump was knackered. All was then made clear as the ever-ingenious Fred ripped the old Hotpoint apart, removed the motor and told Bill to dump the rest back with the scrapper, adding: 'There's enough crap around the yard as it is!'

Clear as mud? Well, of course not. Fred built a stone-cutting lathe and used the washing machine motor to drive it. Bill recalls that the stone was bought from a local firm, Bramhalls Monumental Masons, and a sample column was made. The pair took it to the powers-that-be for approval, which was readily given, and the rest of the columns were made to the same pattern. Scaffold was erected on the town hall roof so work could commence.

Bill recalls that the job went well and, as always, Fred's work was of the highest standard. But for the colour of the new mortar, you had a job to tell the new from the originals, he declared.

During a smoke break they had been looking around the dome and noticed that what they first thought were little blobs of stone above their heads were, in fact, lions' heads cut into the stone. Fred was preoccupied with this for some time.

The last day of the job arrived.

'We had hauled all the last gear up to the scaffold including a new batch of mortar which we kept loose with water we had brought up with us,' said Bill. 'When we had finished we both lit up and stood back to look at out handiwork. "Champion", said Fred, and I have to say the columns really looked the business.'

It was then, Bill recalled, that Fred's face acquired a cheeky smirk as he gazed at the stone lions' heads above them.

'I am going to make them buggers see,' he declared, as he pointed to the carved lions. He then produced a handful of glass marbles like the ones kids play with from his pocket.

No sooner said than done – and up he went. His drill reamed out a socket in each of the lions' eyes.

The pair hadn't noticed the impending gloom and as they began to mortar the 'alleys' into their new homes, they became aware that the mix was getting too dry to work with.

Bill recalls what happened next.

'We realised that we would not be able to go down for water – a precarious round trip of about forty minutes – as by the time we got back it would be pitch black. A Victorian town hall roof is really no place to be after dark.

Crisis Over!

'There was nothing for it. Needs must when the devil drives, as they say.

'Having worked up there for four hours since our last break, we both, fortunately, had very full bladders. The strained best bitter worked perfectly in conjunction with the mortar!

'We finished the extra job, thus becoming immortalised in urine at the top of Bolton Town Hall.

'Not a lot of people know that!'

Fred's Way

Fred Dibnah almost certainly started working on old buildings and factory chimneys out of admiration for the skills of their creators, but unfortunately he was born too late to have any lasting effect on their salvation.

Traditional industries were starting to show signs of age and the industrial Britain which Fred loved was undergoing big irreversible changes. The demise of those traditional factories and in many cases the redundancy of their very fabric meant that Fred's embryonic enterprise would find work. However, those labours were not, in the main, to be concerned with careful maintenance but in planned demolition.

The well built traditional boiler house chimneys were creations of great beauty to Fred and he carried out significant repairs to a number of them, some of which have survived

Fred was always happy to show people around the yard. This is a group of youngsters from Tyldesley primary school who visited in March 1980.

Fred with his autobiography in December 1983.

into this new millennium. In Fred's eyes, the structures were monuments to the people who built them, artisans he often referred to as being 'hard men'. To destroy their work was bad enough to him – but to do it with dynamite was something he saw as insulting, and 'not fitting'.

He started his chimney demolition career by applying what he called 'the science of back'ards construction'.

Fred made sure that this redundant factory chimney in Eccles had a good last smoke in August 1991.

Having climbed the condemned chimney, Fred removed the bricks course by course and either dropped or lowered them safely to the ground. There was another way which got the job done, no explosive charge was employed but plenty of drama was created. Because of the relative slowness of the method and the fact that it allowed the chimney one last smoke, in Fred's book it was a more fitting end!

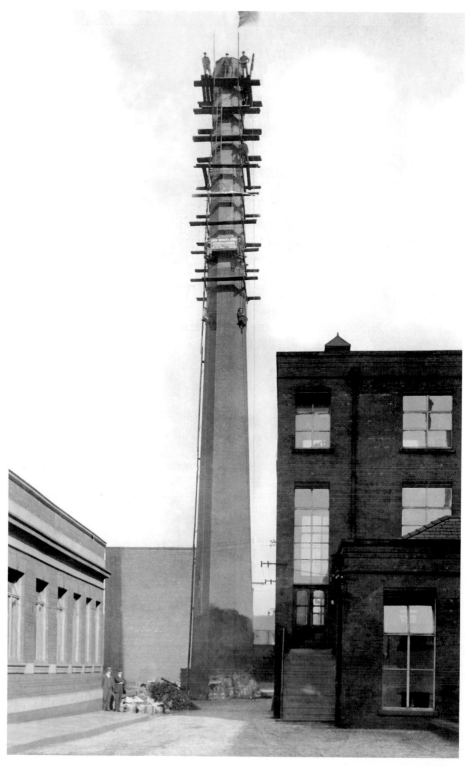

Fred always enthused about the high class workmanship of the old 'hard men', as he called them. Some of those builders are depicted in this 1905 picture of the newly completed chimney at Richard Hardwoods Ltd.

Fred is pictured cutting out the bricks on the side of the chimney. He is being assisted by Neil Carney. Warrington, 1994.

Simon Warner's first chimney drop. The fire is lit. Warrington, April 1994.

A close-up of the props put in to hold up the chimney temporarily.

Down she goes! A 150ft chimney at McKechnies Copper Smelting Works in Widnes bites the dust. March 1991.

Fred made this method his own, turning each toppling into an enthralling piece of live theatre. If you doubt it, just ask anyone lucky enough to have been there.

On the face of it, Fred's tried and tested method was simplicity itself, although the danger was very real. Only by applying the skills born of many years' experience did Bolton's most popular son make it look so easy.

Having decided in which direction they wanted the redundant structure to fall, the demolition gang first cleared a site, by necessity a good bit larger than what Fred had decreed would be the chimney's 'lying down' size.

Then Fred and his team took over.

Slots were cut into the side of the chimney and the bricks were removed from the opening, usually making it approximately three foot high by one-and-a-half foot wide – or, if you prefer, two bricks wide by twelve bricks high (apologies to those of a metric leaning). Fred measured his work using an ancient yardstick. Two sections of scaffolding plank (approximately eight inches wide) were then cut so that their length just exceeded the thickness of the bricks in the chimney wall.

The next move was to insert the 'filling' into the scaffolding plank sandwich, a piece of wooden ex-telegraph pole or similar, which took on the role of a 'pit prop' and so held up the bricks above it, taking the place of those which had been removed. After leaving a width of about three bricks, the exercise was repeated further around the base of the chimney until 10 or so such props were in place.

The bricks between the props were then carefully removed so that the chimney, on the side chosen for the direction of the drop, was entirely supported by the newly inserted wooden supports.

Then, with the aid of an electric drill – although in the early days Fred used a good old chest brace – a couple of holes are drilled through the width of each pit prop. The holes, Fred used to explain, were put there as the result of an observation during an earlier chimney drop and their purpose becomes obvious.

The next essential ingredient was timber debris from surrounding demolished buildings. Selected sections were packed between the props from the bottom to the top of the 'letter box' shaped slot. That done, a huge bonfire of the remaining scrap wood was built against the side of the chimney, completely covering the props and the inserted wood. Several gallons of diesel fuel were added until all the wood was well and truly soaked.

All that was then required to topple the chimney was a piece of wood approximately two inches in length and one eighth of an inch in cross section. Provided the brimstone covering on one end of that piece of wood was nice and dry, a quick application of the laws governing friction did the rest.

Why the holes in the props? Fred's earlier observation concerned a split prop. He saw combustion being aided by flames getting inside the wood, resulting in a quicker burn rate.

As the fire got hold, the supports would burn away and Mr Isaac Newton's law did the rest! There is nothing more to say on the subject except perhaps, 'Do you like that?'

Fred explaining, as only he could, why having a mine shaft and winding gear in his back garden was a perfectly logical thing to do!

The Man from the Telly

With the great gift of hindsight it is easy to see now why Fred Dibnah became a TV personality. He was, as has been said many times before, an absolute natural.

The first series of films featuring Fred was made by Don Howarth for the BBC during the 1980s and grabbed the nation's imagination. Because of that initial success it was perhaps inevitable that others would follow.

Fred at the top of Bury parish church. This is a still from the film clip which caused all the trouble with the Health and Safety Executive.

Alistair Macdonald reaches the top of Bury parish church to interview Fred, 1982.

Going: The demolition of a chimney at Farnworth, near Bolton in 1991.

Gone: The demolition of a chimney at Farnworth, near Bolton in 1991.

After following Fred to the top of the chimney, Alistair realised that they were sitting on a 200ft high pile of loose bricks!

The familiar face of Fred was still to be seen regularly on British TV as the new millennium dawned, and the advent of satellite broadcasting has now introduced our hero from Bolton to an even wider audience – even though satellite dishes were not high on Fred's list of pleasing modern artefacts. In fact, he hated them and often referred to them as 'woks on chimneys'.

The programme which started it all was first aired in 1979 and simply called *Fred Dibnah – Steeplejack*. It was an hour-long special, part of a series about people with unusual jobs. It was commissioned following a news item on the then *Look North* evening regional programme for the BBC in Manchester – but in fact Fred had appeared on TV a year or so earlier.

A young BBC reporter, Alistair Macdonald, was sent to climb to the top of the town hall in Bolton to interview 'a steeplejack'. He cannot have imagined as he climbed what an amazing phenomenon he was going to be responsible for discovering.

After that memorable day, Fred Dibnah self-employed steeplejack, would no longer be a private individual living and working in the north-west of England. The young man from 'aunty' had started Fred, albeit unwittingly, on the road to becoming Dr Fred Dibnah MBE, and consequently a national treasure.

Having had a successful career of more than twenty years in BBC TV news reporting, Alistair Mac (as Fred referred to him) spent a short spell with the other side (ITV) and then went on to make award-winning documentary programmes.

He left broadcasting to set up his own company, AMTV, which specialises in honing the communication skills of business people. He is also involved in teaching general presentation skills and he is a sought-after keynote speaker and conference chairman. He is also the managing director of the Arcturus Training Network Ltd.

On the face of it, the 'Beeb' could not have chosen a more suitable interviewer, as Mr Macdonald was then – and still is – a keen mountaineer and hill-walker. However, perhaps the word 'chosen' is not the right one as we discovered when we visited Alistair at his delightful Derbyshire home.

What follows is his amusing account of the time when, unwittingly, the BBC discovered Fred. We also were somewhat surprised to discover why there is, to this day, a particular clip of film featuring Fred, which is still very important to the BBC.

Alistair Macdonald recalls:

In 1978 when I was working as a reporter on *Look North* as it was called then (now *North West Tonight*) the news editor said one morning: 'There's some nutter of a steeplejack regilding Bolton Town Hall. Does anyone fancy going up there to interview him?'

His emphasis was very definitely on 'up there' being the top of the town hall and not just a visit to Bolton. That fact dawned on all of my colleagues at the same time and as a consequence you could have got knocked over in the rush as they hastily left their seats and headed for the canteen!

I was still at my desk and I noticed the editor's gaze fixed firmly on me.

'Alistair, you're a climber. Do you fancy doing it?' he said.

Much to the relief of my reluctant co-reporters, I agreed to go. I did like to rock climb in my spare time and this assignment sounded as if it would be fun.

The British Hardwear and Tools Ltd chimney pictured in 1968 after it had received Fred's attention.

The view from the top of Atlas Mill, the kind of vista Alistair and Fred saw as they put the world to rights.

I went to Bolton with a cameraman called Neil Higginson. We had been warned that it was all likely to be a bit precarious and that the council was showing some concern about the health and safety implications. As I learned later, health and safety was an area of almost no concern whatsoever to Fred.

Neil and I went up through the of Bolton Town Hall (inside staircases) and then climbed out through one of the pillars onto something I can only describe as very rudimentary staging. We stood a moment or two in order to get our balance and then, treading very carefully on the somewhat rickety scaffolding, moved to where we could see Fred high above us, busy at work regilding the dome, a job I believe he was doing for next to nothing because he wanted his name there, on the dome, for posterity.

We stayed up there long enough to get enough film to make an interesting little piece for *Look North*, which went out later that day. It was, in fact, very good and showed Fred at his best. Don't forget it was his first TV programme but he really was terrific.

We returned to the newsroom the next day and everyone was raving about our Bolton Town Hall film. Not only did the public love it, but some of the BBC bosses were also very impressed.

As a result, accompanied by a couple of my colleagues I went to see the head of local programmes, at the time a guy called Ray Colley.

We all told him about how brilliant this guy from Bolton had been and what a great character he was. Now Ray was a top flight TV executive and he had a reputation for having an unerring eye for what would or would not make a good programme. He had also seen the news piece – but this was not, as things turned out, to be one of his better days.

To bring the discussion to what I hoped would be a decision-making conclusion I said, 'I think this guy and his work will make a great half hour documentary.'

Ray Colley then famously (that is in BBC circles) made what for him was an extremely rare misjudgement. Sitting back in his big chair, he slowly took his ivory cigarette holder from his mouth and then replied: 'I doubt if it would make ten minutes.'

The answer was no, on this occasion.

We went away with our tails between our legs. Being quite new to broadcasting we wondered if we had got it wrong.

As luck would have it, Don Howarth saw the news item later and he immediately saw the potential – and that is how the first series came into being.

Fred always when we met hailed me as the 'guy who got me into TV' which, technically, I probably did, but at the end of the day he did it all himself.

We remained friends after that first meeting at the town hall. I went to other chimney events and visited him and his family on several occasions. The welcome was always the same, warm and genuinely friendly.

I recall seeing Fred's Land-Rover heading up the street towards me one rainy day while I was walking through Manchester city centre. He spotted me at the same time and just stopped there and then for a chat. As it was raining his mate jumped into the back of the jeep on top of some ropes and I edged into the front passenger seat.

We were happily yacking away twenty to the dozen when there came a 'rat tat' on the window. There, in the rain, was a traffic warden who instantly recognised Fred and I think knew me, too, from my appearances on *Look North*.

She pointed out that we were on double yellow lines and asked ever so politely if we would like to continue our chat somewhere else. She was so taken with meeting Fred that she directed us to a 'safe place' to park out of the way, which I think was in a side street at the back of a big department store. Such was the magic appeal of Fred.

Fred liked to test his friends as others have related and I was no exception.

I got a phone call about a couple of years after we first met and it was by the way of an invitation, or maybe a command!

'Do you think you could go up a bloody chimney – you know, straight up?' he asked.

It transpired that he had laddered the chimney of the destructor plant in Bolton, reputedly the second highest chimney still standing in Lancashire. 'Come over Sunday and see if you can climb it, then we'll 'ave some lunch.'

So, as requested – nay, instructed – I turned up at about 10am. I was full of trepidation as rock climbing and ladder climbing are totally different things. I really do not like climbing ladders.

Fred set off and after a while I followed him onto the rungs. It was very hard work but as always, Fred made light of it. He had got to the top and was sitting on the rim of the chimney. As I got to the top I saw to my horror that the ladder stopped about four feet short of the chimney lip and there was no staging.

'Just reach up and pull yourself up,' Fred said, 'You'll be right!'

I was right, right worried, it was a hell of a thing to do and on a vertical surface, not a sloping cliff face.

We sat there, putting the world to rights, as they say, but I was preoccupied studying the parlous state of the chimney. There was, it seemed, no mortar. Hell, I thought, we are sitting on a 200-odd foot high pile of loose bricks!

Fred was pointing out landmarks and this and that but all I could think about was how I was going to get back down to that ladder?

I mentioned to him that it was getting on for a quarter to two. In those days the pubs closed at two on Sunday lunchtime.

'No worries on that score, cock,' he said. He had that sorted out.

'You go first,' he continued.

I, very gingerly, lowered myself back down to the ladder.

Even though I dislike ladders I was very glad to feel my feet land firmly on that one. Down we went and then off to some pub where Fred said time was not an issue. It was not, and we had more than a few pints. Eventually the landlord decided that he wanted his Sunday dinner (lunch for those of a southern persuasion) and we were all quietly ushered through the back door.

Preparing to drop the mighty Atlas Mill chimney in Bolton.

After Fred's fire had done its job, the laws of gravity took over at Park Mill, Oldham in 1985.

I thought we would then go home to Fred's for lunch, but he was not yet ready. We carried on supping, this time in what I think was the convivial atmosphere of the Ukrainian Club, but we were not drinking beer.

Fred wanted me to try some honey liqueur. It was pure rocket fuel and after a couple or three I was flying. I had to feel my head to tell if it was still attached and also try to straighten my blurred vision. On that occasion we were more than a little late for lunch!

Some time after, Fred called and asked if we would like to join him for, as he put it, 'a bit of telly' at the top of Bury parish church. We got the OK from our producer and Neil Higginson and I and, I think, a soundman, set off for Bury.

Fred was fixing a new weather vane to the church steeple and as I recall it was a fairly hairy climb in what felt like a force 10 gale, although it was a nice bright day. At the top of the spire, Fred had in place a very skimpy-looking scaffolding rig, with a couple of planks on one side and a single plank on the other. We got what I recall was a great piece of film. By this time Fred was a seasoned broadcaster having completed his first and possibly his second TV series.

The interview was done with him leaning out over the town below, nonchalantly smoking a fag and with one arm crooked around a scaffolding tube.

It was then that I asked him the daftest question of all time.

'Have you ever had an accident, Fred?'

Of course, he came back with the classic one liner.

'You only have one accident in this game and then it's half-a-day out with the local undertaker.'

That particular piece of filming will always stick out in my mind because of what happened the day after it went out on *North West Tonight*.

All hell broke loose at the BBC. In fact, the shit hit the fan in a major way.

The Health and Safety Act had just come in and the inspectors turned up at Oxford Road studios, demanding not only to see the film which had been broadcast the night before, but all the off-cuts as well.

While they were being shown to an office at the back of the building and plied with tea and biscuits, the off-cuts mysteriously disappeared, but they were not to be outdone. There was a new set of regulations which they were going to enforce, come what may.

Fred and Alistair Macdonald, seen together at Radcliffe Road.

They had seen the broadcast and that was enough for them. They pointed out that not one of our crew was wearing a hard hat and, furthermore, Fred's scaffolding did not comply with any of the regulations. Fred also had no hard hat and so on. Their list of misdemeanours seemed endless. They sent a separate inspector around to see Fred. I was not there but would like to have been.

Apparently, when they told him about failing to wear a hard hat, he said: 'Well, when you fall from 200 feet, it's no bloody use having a hard hat on.'

They finished both sets of interviews and then informed both Fred and the BBC that they would be taking out a prosecution under the new Act. We were both actually looking forward to a day in court but they dropped the charges and it never happened – but it did have major repercussions for the BBC.

Months later we had all forgotten about the Fred incident, but nationwide, matters health and safety were getting higher by the day on everyone's agenda. That included of course the dear old BBC. The head of health and safety from London was due to visit Manchester, and, I suppose, all the other regions in turn.

A date was set and at the appointed time we all assembled to meet him and his staff. He delivered his talk and we were all issued with the appropriate leaflets and information.

We listened, nodded and smiled in all the right places. He was very convivial but firm and to the point.

Next up was a film showing various aspects of the application of the rules regarding the new legislation. We all settled down to watch.

Imagine our surprise – nay, shock and horror – the first 'What Not to Do' film shown was ours, made with Fred on the church steeple at Bury. I don't think the health and safety guy realised why we were all falling about laughing at what, to him, was a deadly serious message!

It stayed the number one 'don't do' at the BBC for more than ten years until it was finally replaced by that famous Noel Edmonds motorcycle incident from the Big Breakfast Show – but it only dropped to number two and I reckon to this day it is still high on the BBC's list of safety 'don'ts'.

Fred had such a strong and vibrant personality that he instantly appealed to all kinds of people, whether in person or on TV.

He also had a very highly developed sense of timing and with that gift he was able to keep the viewer fascinated and enthralled even as he explained the most complicated of things.

He did it in their familiar language and at a comfortable pace, never using any gobbledegook or techno jargon. I would say that he was unique.

I am immensely proud of the small part that I was privileged to have played in launching what became a distinguished career in television spanning more than thirty years. I also valued greatly and benefited from the friendship of Fred and his family.

Where's Me Ten Bob?

Bill Richards: Licensed victualler, mining historian and aircraft mechanic

Like others of Fred's close acquaintance, Bill Richards is an ex-coal miner and, perhaps more importantly, he is a diligent and ardent collector of artefacts and records connected with the history of that once-great industry. Bill is also involved with engineering restoration and, like a great many who once earned their living underground, he is fascinated by the art of flight. Traditionally, miners have been linked with the flying of racing pigeons but Bill Richards takes his flying a bit further – he assists in the rebuilding of WWII aircraft engines.

Bill is to be found these days behind the bar of the 'Greenfields' private members' club in Westhoughton, near Wigan, which he runs with the help of his wife, Angela, and his family. In the early 1990s he would still have been found driving tunnels underground, working with a tunnelling contractor. For thirty-one years before that, he toiled at various coal faces far below the county of Lancashire, a part of the world which he dearly loves.

Bill worked as a contractor on the very last tunnel excavated at Parkside Colliery, a heading which opened on to a 'virgin ground' coal face 220 to 250 yards long and which was estimated to have enough quality coal within it to last until 2017, and possibly beyond. There were millions of tons, of which not one saleable shovelful ever reached the surface. The so-called 'modern pit' was sunk in the 1950s and was closed – many say prematurely – after a bitter dispute in 1993. That followed a brave battle in which the miners and their families, with the backing of the NUM, tried in vain to keep the facility open.

Bill needed little persuasion to talk about his friend of many years and how he first became involved with Fred – not, as you might think, in a licensee-and-customer relationship. It was long before that and involved the age-old skill of riveting. This is Bill Richards' tale:

Like a great many friendships, ours started out of need and really at the suggestion of a third party. A mate told me, one night over a pint, that this steeplejack he knew was restoring an old steam roller, and his account of what was occurring, not many miles away from my own front door, fascinated me.

A few nights later the same guy called me over as I was having a pint after work and asked me if we still had blacksmiths working at the pithead. I told him that we did, but added 'only just about' as things were looking grim and we were all expecting to be finished up sooner than later. He nodded and said thanks, nothing more was mentioned, and we both returned to our friends.

A week or so later, I was again in the same pub, when my new-found friend came through the door and spotted me. He started waving a piece of paper and gesturing that I should meet him at a quiet corner of the bar. "Do you think the blacksmith's shop will have any of these?" he said, pushing a grubby note under my nose.

That was the first time, albeit unknowingly, that I had ever set eyes upon Fred Dibnah's classic writing style.

There were other signs of Fred on that piece of paper - an oily thumb and fingerprint. It turned out that the guy with the steam engine, who I was told lived in Darcy Lever, was in urgent need of the items sketched on the paper and conforming to the dimensions as noted. It was a request for a stock of good. old-fashioned rivet blanks.

Bill Richards with the coal mine pump.

I was later to learn that they were urgently needed to complete the boiler rebuild Fred was undertaking at the time on a steam roller he'd purchased a little time before. I went to the pit the next day and, before going underground, I paid a visit to the blacksmith's shop. The 'smith' was a great man of the old school, and he was very interested in what the rivets were required for, so I explained about this guy in Bolton and his old steam roller.

After I finished my shift I called to see if he had found any of the rivets Fred had asked for. The old guy had gone home, but by the door of his workshop were two hessian sacks tied and one was labelled 'Bill Richards x2'. I picked up the note and read the words: 'Bill, these 'ul do 'im, 10/- to pay'. There was a minor problem: if you have ever tried to lift a sack of rivets single-handed into the boot of a car, you will know what I mean. They are bloody heavy! There was also a very neat wooden box stacked alongside the bags with the words 'an' this' chalked on it.

Anyway, by following the directions given to me, I went to meet the bloke with the steamer. As you would expect, he was lying under the thing working away, and he slowly crawled out as I shouted a greeting. I showed him the rivets in my boot and he was over the moon, and between us we soon had then stowed safely away in his shed. I was fascinated with the engine and, as you all know, if there is one thing Fred liked it was a receptive audience for the tales of his work. I was definitely that and we got on famously and, of course, went on to remain the closest of friends.

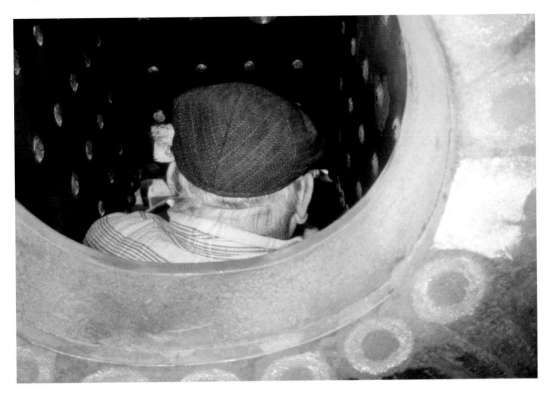

Fred inside the firebox, riveting.

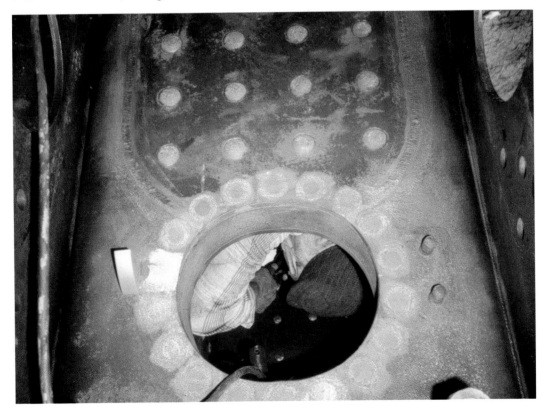

Maker's name on the mining pump.

September 2003 and the convertible begins to take shape. Note the specially laid railway tracks to facilitate movement.

Fred working on the Aveling firebox.

Mine host. Bill at the bar of the Greenfields private members' club.

But that is not quite all about that meeting. There was the matter of the wooden box which was, in fact, hand-made and looked more like a posh presentation case. I was to learn, by watching him open it, that it contained a brand-new set of air-operated rivet guns, one large and one small, with a full set of 'dollies' (rivet head shaping tools) for each, which on first seeing them caused him, for once, to go speechless. He was like the proverbial dog with two thingies and I reckon he rated those guns back then higher that he would have done the Crown Jewels.

That was back in the late 1960s and Fred went on to use those guns for the rest of his life, a good ten shillings'-worth. Therein lies a tale. As I left the yard that night, he thanked me for the gear and, as I headed off, I remembered the bit of a label on which the blacksmith had written the price, so I walked back to him.

'Fred,' I said, 'the old guy wants ten shillings for the stuff and I will have to pay him when I'm next on shift.'

'Fair enough,' he said. 'It's a bloody bargain, cock. The guns alone are worth twenty times that.' Then there was a slight half-smile as he dug into to his overall pocket: 'Got no brass on me. You'll be down the yard again; I'll gin it ya then!'

I did go to the yard again. In fact I went many hundreds of times and, right to the sad, premature end of his life, he joked about that bargain we rescued from a pit that was about to close. It was a bargain by anyone's standards and especially for Fred. I paid the blacksmith, and I don't think for a minute that he, in turn, paid the NCB, but dear old Fred never did get around to giving me that 'ten bob'.

One of my main interests is aircraft engines and, in particular, examples from the time of WWII, and Fred was very interested in this work. We were planning to get together and take a restored Rolls-Royce 'Griffin' aero engine to one of the vintage shows this year (2005) and display it alongside his newly-restored convertible. Sadly, Fred will not now see that happen but maybe it's a project his boys, Jack and Roger, can consider.

Our little engineering society has currently two WWII 'Merlins' on the go and they both came from bomber aircraft. However, our priority project is the restoration of a 1930s-designed Rolls-Royce 'Griffin' engine that last saw service in an RAF Shackleton and was retired only in 1990. I am keen to talk to anyone who is carrying out similar work, as these engines (outside a preserved aircraft, of course) are very rare. The 'Griffin' is a 37-litre power plant as opposed to the 'Merlin', which is only 26 litres so, when fitted to Spitfires, they obviously made the fighters more powerful.

Fred was always keen to talk about aero engines but being, as we know, the all-round engineer, that is not surprising. Another favourite subject of his was mining and, as his last TV series confirms, he liked the company and conversation of ex-miners. On the many occasions he came down to the club for a 'jar of the black stuff' the conversation would often turn to a discussion about the authentic miners' lamps we have around the place and out would come the old mining pictures.

There is one 'Fred' project that I hope will still go ahead, and that is the restoration of a diesel-driven mine water-pumping engine, a very rare one that we have stored at Greenfields, and which Fred was planning to restore. I hope now that the Bolton Mill Engine Preservation Society will be able to fit it into their busy programme.

Fred did so much for the industrial preservation movement so I hope others will keep up that very important work in his memory. Like a good many, I greatly valued his comradeship and I am thankful of having so many memories of the time I spent with him. He will be greatly missed.

That Reminds Me!

Being reminded of Fred's exploits at the Last Drop village Bill Richards to recall a rather similar amusing incident.

Bill is the owner of Greenfields Private Members' Club in Westhoughton, a pleasant watering hole which was often frequented by Fred. So close was their association that after Fred's death, Bill continued to work diligently on his pal's behalf having been appointed chief executor to Fred's will.

Fred's beloved roller pictured outside his home on Tuesday 16 November 2004, the day of his funeral.

Bill with Fred and Alf seen after the pit head gear had been built in 2003.

Fred working at his forge.

Having been introduced to the mystic art of steam preservation, Bill started to visit his new found friend regularly and took a great interest in the progress of Fred's steam roller project.

Some time after the roller was finished, the two men were one day chatting in the yard as Fred waited for the Aveling to get up to working pressure. He had recently corked some steam leaks, said Bill, and, being pleased with the outcome, he proudly pronounced the roller as being 'fit for owt!'

'Fancy a bit of a trundle?' Fred asked while nodding at the engine and making that well-known male sign of a pint being lifted to the lips with his right hand. Bill did not need asking twice and was up on the footplate next to Fred in a trice. He was instructed by Fred to steer.

'I'll drive and you watch and learn so you can do it on the way back,' said Fred.

Off they went, their destination being the Lever Bridge Arms, a popular hostelry a little down the road – the important word being down.

It was wintertime, Bill recalled and the midday sun had hardly managed to break through the grey sky, so there was still a good nip in the air. All was well and the roller performed a

The joy on Fred's face is evident as his winding gear is worked for the first time. Bill Richards is operating the control valve.

treat with the boiler doing everything its designer intended it to do. The boys were in good spirits, says Bill, especially at the prospect of a few scoops in the 'Bridge'. It is worth mentioning that the road to the pub is a gradually falling gradient – that is, until you get to the last bit just before the pub, when it becomes a steep slope.

As the boys rounded the final bend and started to go down towards the pub, Bill recalls that all was not well. It suddenly dawned on him that the roller was developing a life of its own!

Fred was frantically winding on the brake and he shouted to Bill: 'Grab the bloody wooden chock and chuck it under the wheel.'

Scrambling down off the footplate – not an easy thing with a slipping ten-ton roller – Bill attempted to do as Fred asked, but not only was there a nip in the air, on that section of road shaded from the weak winter sunshine there was also a layer of black ice!

Bill got the chock under the offside back wheel twice and on each occasion the roller simply jumped noisily over it.

'Do it again, do it again!' Fred repeated in what Bill described as an agitated voice. The air, he said, was not only white with steam and black with smoke as Fred had managed to get the roller into reverse – it was, as the saying goes, blue!

Alarmingly, by this time the roller was almost sideways onto the approaching bend when they both at the same time spotted a green GPO Morris 1000 van coming the other way at speed.

The driver reacted quickly and managed to get partly up onto the pavement and miraculously squeezed by. Bill says that he is sure that the phone engineer was a good Catholic for as he shot by he was frantically making the sign of the cross! The waltzing roller sailed on, past the pub's smoke room window. Somehow Fred had managed to slow her enough to gain entrance to the car park at the side. Bill said that they stopped about six inches from a very solid wall. He added that both men were definitely on the verge of becoming brown trousered!

The landlord of the pub was a very taciturn sort of a guy and rarely given to humour, Bill recalls.

As they walked in, Fred was first to the bar and the shock must have had some sort of effect on him as he had folding money in his hand! He gasped: 'Give us two pints, cock.'

The gaffer calmly started to pull the ale and, without smiling or looking up, simply said: 'Thowt tha wa' comin' in, seed thee goo past!'

As the rest of the guys in the bar collapsed with laughter he looked up and winked broadly.

I had better not say what state the two were in when they piloted the roller back some three hours later, but I can confirm that they were on the lookout for GPO vans!

d Dibnah in pencils, drawn by his friend Bill Greenhalgh.

The Dream?

Bill Greenhalgh has turned up a copy of one of Fred's drawings which simply poses a question: Did Fred always dream of having a pit head of his own and a workshop with a distinctively designed chimney?

   The pencil sketch drawn by Fred in his own inimitable style was dated and signed by him on 25 November 1960. He was then twenty-two years old and it was drawn a full seven years before he went to live in the Victorian gatehouse which became his home in Radcliffe Road, The Haulgh.

   In the 2003 colour picture, Fred is seen with his friend Jimmy Crooks as they worked on the mine shaft. Note the shape of the chimney stack above them. It is very similar in style to one to the left of the laddered factory chimney in Fred's sketch. The similarities do not end there as can be seen by examining pictures of Fred's replica pit head gear!

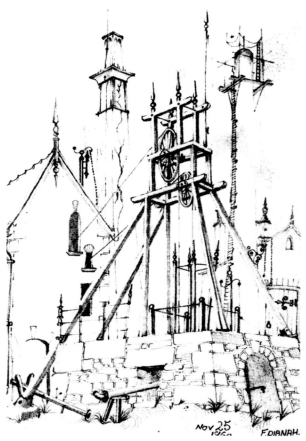

Above: Fred's pencil drawing done in November 1960. Note the shape of the smaller chimney (to the left) and compare it with his actual workshop chimney in the other pictures.

Right: Fred and Jimmy Crooks giving the job a coat of looking over, 2003.

Above: Fred's living van is seen alongside the pit head gear in 2004.

Left: Fred's pit head gear seen from another angle.

Fred explaining the basic principle of the pit head gear in his garden at Radcliffe Road in 2004.

'Up & Under'
Simon Warner Remembers

Expertise, strength of character and ingenuity are qualities that seem to be found in abundance among Fred's friends and colleagues. Add flair, imagination, loyalty and a good sense of humour and you have an all-round description of the unique band of people who were Fred's real pals. Simon Warner was proud to be counted in that select number.

When you are made aware that Simon's business logo slogan is 'Up & Under', you may justifiably wonder why. However, it's easy to understand when you know that he is both a steeplejack and a gravedigger. The later calling came when his dad (now retired) was a vicar so, as a young man, Simon saw a lot of graves, and as for the former, he was taught by, and worked with, Fred.

Simon takes up the story:

After leaving school in 1982 I went to Bury Technical College to continue with further education and that was by doing an engineering course. Living in the Bolton area, I was aware of this guy called Fred Dibnah who 'did steeplejacking work' and his involvement

The end of Asia Mill chimney in November 1998. Most of the props are in place.

Fred relies on the old-fashioned hammer while the 'sponsored' nail gun can be seen swinging on the rope below him. He couldn't master its double safety catch mechanism. Simon Warner (left) and Fred on Badger Homes' chimney at Chorley.

Repair underway. Note the newly-seated band.

with that activity caught and held my attention, as the subject had always interested me greatly – so much so, that I plucked up the courage and went to see Fred one evening to discuss the possibility of me working with him in some capacity.

He was very nice to me but tried to dissuade me, by pointing out that most of the mills had been demolished and saying that 'there will soon be no chimneys left to repair'. He was probably right, I thought, but surely there were still many churches where his services were required?

At the time I was disappointed that nothing positive came from our conversation. His advice was to stick to the engineering and gravedigging. (I was gravedigger at Stand Church, Whitefield where my father was Rector). Even though I was disappointed at the outcome of my visit, I still had the eager longing to work with him in his most unusual occupation. This thought never left me.

Some ten years later, while at Bolton Institute and then working for an engineering degree, I saw Fred's TV programme. It spurred me on to contact him again. On this visit I arrived on my 1971 Triumph 650cc Tiger motorcycle, knowing that Fred also had an old cherished machine, an AJS motorcycle. Fred was doing something, which I later became aware was a regular thing: he was showing a group around the workshop and sheds which housed his steam engines and other equipment.

I tagged on to this group; and, when they left, I stayed behind. I reminded him of my previous visit and said that I still had hopes of being able to work with him. It turned out that Fred was due to fell a chimney in Warrington the next week and he encouragingly suggested that it might be possible for me to assist in some small way. I gave him my name, address and phone number in the hope that I would hear from him.

Fred swings in a bosun's chair high above the housing estate.

Simon and Fred pose on the top of a chimney. Note the lightning conductor tied to the ladder and about to be fixed to the structure.

I was already climbing at this time and worked in the film industry as a rigger, fixing lights and sound gear etc, and that was something I really enjoyed. I had passed my Heavy Goods Vehicle licence but I still dug graves when called on to do so.

However, more then anything else I still wanted to do some steeplejacking with Fred.

This time the phone did ring and Fred invited me to be at his house early on the following Monday morning and so join him on a job. I was there in good time and turned up at Fred's ready and eager to help. We worked hard all week getting the chimney ready for dropping and, come the following Sunday, all was ready. Fred gave me the honour of sounding the horn that indicated the chimney was ready to fall.

My parents were among the big crowd who had turned out to see this event and later told me that the experience of being so close to this very high chimney as it was about to collapse left me as white as a sheet! It was, as anyone who has been close to one will confirm, an unforgettable experience. That week was the start of a seven-year working association with Fred and the experiences are counted as being among the best in my life so far.

I worked for him on and off, on a part-time basis, until I finished my studies at Bolton Institute, after which I then worked with him almost full-time until 1999.

ome of the typical damage that Fred and his team were
lled upon to repair.

A job well done. Fred sits back to admire his work.

This period was a steep learning curve for me, as Fred taught me to become a competent steeplejack. He cultivated the laid-back, devil-may-care persona, but let me assure you: when he was climbing no risks were ever taken. All the gear was checked out and nothing was left to chance. On any job we all knew what we were each to do and when we were to do it.

Interestingly, Fred would not climb on ladders I had fixed to a chimney for about two years; it took him that long to be sure I knew exactly what I was doing.

Our very interesting work took us all over the UK, up church towers and spires and repairing and demolishing chimneys. There are a great many stories to be told and the TV coverage of Fred, even as comprehensive as it is, shows only a small part of his challenging and fascinating working life.

The story of the Canvey Island chimney is told elsewhere and I am sure you will find it both interesting and amusing, but there is one tale I must tell, as the picture to accompany it is priceless – but then, most pictures of him are!

We will describe this company as a famous firm who produce power tools specifically designed for shooting fixings and nails into all kinds of material, including brick. The equipment replaces the need for conventional hammers, and therefore eliminates the difficulty in using them with one hand holding the nail and the other the hammer, ie while climbing. At least that was the theory, but it didn't apply in Fred's case.

Brew time at altitude. Simon Warner with Fred.   Eddie Chatwood, a regular member of Fred's team.

We were booked to drop a chimney for the construction company Badger Homes in Chorley and, in order to publicise their operation, the company asked for a banner to be nailed on to the condemned chimney, a feature which is quite normal and often requested by the client. In this case it was sort of 'double bubble' as the nail gun manufacturers had asked Fred if we could use their latest product to fix the banner. They had a photographer on hand, telephoto lens and all, with which to get stunning publicity shots. Good plan!

So up we go, one of us swinging on each side of the stack and both of us gunning the fabric to the bricks as we slowly worked around the edge of the banner. Or at least, I thought that was what was happening. I became aware that Fred was going a bit slower than me and then realised why. He was using a traditional hammer and nails; the state-of-the-art gun was hanging out of use on a rope below him.

So much for the promotional work. He had apparently failed to master the double safety catch mechanism on the tool and had given up and gone over to the tried-and-trusted old way – a perfect example of how he hated modern technology and confirmation that he only really trusted traditional methods.

Fred's other work involved making weather vanes, and repairing and making components for static and steam engines. All Fred's workshop machinery was steam driven

Fred climbing a church steeple high above Preston, Lancashire.

The engine at Wetheridge Pottery.

Barrow Bridge Mill, Bolton. Fred with the lightning conductor.

Fred working on the pulley system at Wetheridge Pottery.

and I was impressed with the accuracy that was achievable with such basic engineering instruments, again when compared with so-called modern engineering technology.

A notable experience of a static engine rebuild, which I worked on with Fred, can be seen at Wetheriggs Pottery near Penrith. The pottery was restored in 1994 and the steam machinery was restored by us in 1995 and was featured in one of the earlier Dibnah programmes. This fine visitor attraction is a pure nineteenth century industrial monument steeped in a history of incredible creative spirit and endeavour, and the fact that it is today the UK's only steam-powered pottery is testimony to that.

When I finished working with Fred, I returned to the entertainment industry as a 'rigger' in arenas and conference centres. This is very interesting work. You get to set up the gear for all the so-called stars of today and we climb using the latest materials and techniques. I even worked on the Manchester Commonwealth Games. But nothing comes even close to my vivid and lasting memories of working with Fred in general and on the chimneys in particular.

I heard the sad news of his death at a time when I was working on a mill chimney in Yorkshire and so had the opportunity to leave my own commemorative mark. Anyone climbing the chimney of that preserved structure, the Lister Mill at Bradford, in the future, will find that I have left a marker in Fred's memory. Like a great many people, I will miss him but I'll never forget him.

Simon left this epitaph to Fred at the top of the Lister Mill chimney in Bradford.

Climbing to Even Greater Heights

After Simon Warner had established himself as one of Fred's team he was eventually awarded what he considered to be a great honour! Having climbed with Fred on a series of repairs in and around Lancashire Simon had still not fulfilled what must at the time have been every Lancashire schoolboys dream, he had never been 'up close and personal' when Fred Dibnah presided over the last smoke of a really big chimney. Not only was he eventually invited as part of Fred's specialist team to a demolition site in Warrington during April 1994 but Simon had been promised by Fred that on the day of the 'topple' he would 'perform a very important act'.

Fred did not really take great pleasure in dropping chimneys, simply because he admired the skills of the craftsmen who had created them so many years ago, and doing so without the benefit of modern cranes and lifting tackle, even the scaffolding they used was crude by comparison with its twenty-first century equivalent. But Fred realised that nothing would stand in the way of 'so called' progress and the big redundant mill chimneys would have to go, in order to make way for new housing and industrial units. Fred's maxim was always 'if a jobs worth doing, then it's worth doing well', and that included demolishing big chimneys!

It was, Simon recalled, 'a big un' and the first day of the job could not come around fast enough for him. So impressed had Fred been with Simon's work on other jobs

Fred puts a giant Comic Relief nose on the 160ft Richard Threlfall chimney, Bolton. 14 March 1991.

and on the preparation for the demise of this monster stack that he had decided to award the young worker a great honour. Simon would be allowed to sound the all important hooter announcing the imminent collapse of the once proud structure!

There was a big crowd gathered on the Sunday of the 'drop', the throng including Simon's parents. As the time for the drop approached, Simon was not found wanting and the horn blared out to good effect. However those watching the event all remarked on the same thing. The young guy on the hooter, they all noticed, had turned as white as a sheet! The chimney fell exactly as it was meant to do, alongside the railway line (WCML near to Warrington Bank Quay station). As for the ghostly pallor, Fred never let Simon ever forget it!

Many examples of Fred's inventiveness have been marvelled at over the years and Simon is able to give us the details of yet another.

While repairing a very tall chimney at Barrowbridge, next to Moss Bank Park, Bolton, the boys needed to haul a large amount of mortar up to the top work station. This was very heavy work and the higher you got with the bucket the heavier it felt, Simon recalled.

'We had all just about had enough when Fred's genius kicked into gear, and saved the day,' he said.

Fred's plan, which worked perfectly, was to rig a pulley and axle on scaffolding tubes at the rear of the faithful old Land-Rover, with a belt drive from that being fitted to the off-side rear wheel of the vehicle. The idea was simplicity itself. Running the engine with the

Simon Warner is seen high above Manchester United's Old Trafford stadium, working on rigging for a Bon Jovi pop concert.

On site at the water tower job at Bolton Royal Infirmary.

A detailed picture showing the way that each corner of the water tower at the hospital had been prepared.

Simon at the top of a repaired chimney at Brightmet, Bolton.

Getting the ornate tower at Bolton Royal Infirmary ready to drop, Fred is pictured in that hat...

Simon Warner and Fred stop work for a chat at altitude.

rear wheel jacked up would drive the pulley. On the other side of the axle was another pulley. Winching a rope around that pulley then allowed the mortar bucket to be hauled up to the job by creating a capstan winch. God knows what the modern health and safety bods would have made of that! What is more, I often wondered how Fred got away with driving the Land-Rover back and forth, which he did for the duration of the job, with the pulleys clearly visible!

Simon had a long and happy working relationship with Fred, whom he worked for in both part and full-time capacities. He finally returned to working full-time in the entertainment rigging industry in 1999. By that time there were not too many chimneys left to work on, he recalled, and Fred, who was nationally known at that time following his earlier TV work, was moving ever closer to developing and expanding his particular niche in the entertainment game.

Above and below show the pulley system. Above: On site at Brightmet, Bolton and Below: In Fred's yard.

Simon with the loaded Land-Rover before setting off for the job at Barrowbridge. Note the pulley. This shows only half the number of ladders used on that job.

Working on the demolition of some outbuildings and what he says was a 'very ornate tower' at Bolton Royal Infirmary, Simon recalls how he and Fred were dogged by a particularly vigilant safety officer as a result of which – and much to his chagrin – Fred had to wear a hard hat and not his trademark flat cap. That did not go down too well!

Where all aspects of his work were concerned, Fred was ultra cautious and very careful, although it may not have seemed that way to the casual onlooker. Simon remembers that he had been putting ladders up the outside of chimneys for over two years before the master steeplejack trusted his work enough to climb on them.

Fred took a professional interest in Simon's work. When the occasion arose the pair visited the firm which manufactured some of the climbing harnesses and safety gear that the young man uses and relies on every working day.

Today, if you know where and when to look, you can see Simon high above the stage sets at venues including the MEM Arena in Manchester and either of that city's football stadiums during outdoor music events. He also worked extensively at the City of Manchester Stadium when that venue hosted the Commonwealth Games.

The next time you, either in person or on TV, catch sight of a pop superstar performing at a live concert, it is worth remembering that everything that is in the air – sound stacks, lighting rigs or sets used during their time on stage – was made possible by the teamwork and skills of guys like Simon Warner. The riggers put in many hours of exacting and athletically challenging work both before and after an event.

Like the man said, there are not that many great industrial chimneys left to climb any more!

Fred breaks into song. "Don't go down the mine Daddy, there's plenty of coal in the shed!"

Fred talks to camera while filming at boltmakers Thomas Smith Ltd in Atherton. Alf casts his eye over the furnace.

Fred and Alf Molyneux with the Aveling Convertible traction engine on Parys Mountain in Anglesey, whilst Roger and Jack Dibnah get ready for the off.

Climbing up to the top of the mountain.

Pouring rain, steam and oil and soot, Alf has copped for the lot!

Chimney Demolition
The Family Man

Born in pre-war Britain during 1938 Fred Dibnah was, according to his mother, always interested in his surroundings and, in particular, anything to do with climbing or that involved steam. As a boy he loved to watch steeplejacks at work on the chimneys and tall buildings of his Lancashire home town of Bolton. From his bedroom window he could see steam locomotives at work, and he would sit and watch them for hours.

While watching those artisans, he studied the way that they erected their ladders and built scaffolding, no doubt formulating in his mind the way he would do the job when his time came. The time did, of course, arrive when, like most people, he needed to choose how to earn a living and support a family. It was to his love of climbing tall structures that he turned in order to be able to put bread on the family table.

When Fred celebrated his twenty-first birthday on 29 April 1959, post-war Britain was preparing to enter the so-called Swinging sixties. Conservative Prime Minister Harold

The first blow is struck: Fred and Mick Barry start drilling another condemned chimney.

The props are in place and Fred checks his calculations.

A nonchalant Fred in 1990 – note the moustache!

MacMillan had a year earlier made the statement he will always be remembered for: 'Let us be frank about it: most of our people have never had it so good.'

But work for Fred, and indeed for many more in the north of England, was not that easy to come by. Life was hard and work challenging, but Fred Dibnah was always up to the task.

The traditional industries were starting to show signs of age and irreversibly the industrial Britain that Fred loved was undergoing changes which, for better or ill, would irrevocably and completely reshape the nation. The demise of those traditional factories and the redundancy of their very fabric was at first a good thing for Fred's growing enterprise. Factory chimneys, which for many years had been carefully built and then lovingly maintained, would now need to be demolished.

The well-built traditional boiler house chimneys were represented in Fred eyes as structures of great beauty, and he often referred to them as each being the greatest compliment anyone could ever pay to the people who built them, those people being workers he justifiably referred to as 'hard men'. To simply destroy them with dynamite at the end of their useful lives did not appeal to Fred Dibnah and anyway, he reasoned, because of their location, many would need to be more carefully dismantled.

This kind of demolition Fred and his colleagues often achieved by what he once termed 'back'ards construction'. Having climbed the condemned chimney, Fred then removed the bricks course by course and either dropped or lowered them safely to the ground. But there was another way that got the job done: no explosive charge was employed but plenty of drama was created. The use of this method Fred made all his own.

From the 1970s onwards Fred's fame as a toppler of chimneys (and tower-type structures) became legendary. Whether it was assisting in the creation of a retail park by removing the control tower from a second World War airfield or the removal of a Victorian chimney for a modern-day housing developer, the system was firstly, dramatically successful and secondly, bound to draw the crowds. Using his tried and trusted method, Fred safely reduced many a once-proud structure to rubble.

Fred's last chimney

The following account is from the occasion of what was destined to become Fred's last chimney. He dropped the structure exactly as planned on Sunday, 9 May 2004.

The way in which Fred dramatically enforced the 'No Smoking' rule for old factory chimneys was without doubt something any observer would remember for ever. Neither big bangs nor a display of aerial pyrotechnics figured in Fred's methodology but the outcome was just as effective and equally dramatic.

Travelling along Manchester's Oldham Road on a quiet Sunday morning can be a leisurely affair, especially if none of the area's senior football teams is at home, so you could be excused for wondering why, on a non-matchday, the crowds were heading towards the old cotton town. Signs announcing street closures were being placed at junctions and the peace and quiet was shattered by the drone of an attendant police helicopter.

What is more, its loudspeaker was warning of the aforementioned road closures. Why? Yes, you guessed it, Fred was in town and he was going to 'knock one down' (chimney that is, not a pint – that came later!) in order to erect even more dwellings to satisfy the appetite of this property-hungry generation. Messrs George Wimpey needed to clear yet another ex-industrial site.

Fred explains to camera what is going to happen.

The Manchester police helicopter circles the doomed chimney.

The older buildings, at Lion Mill in Royton, had been conventionally demolished but the 100-year-old, 300ft stack remained intact. What is more, being within 50ft of the still-in-use portions of the mill, its demise could obviously only be achieved by employing skills of a very special kind. The builders had already started redeveloping the site, on which there are existing residential properties, so no explosives was very obviously the order of the day.

County Demolition Ltd, a Manchester company, knew just what was needed or, more accurately, just who was needed. The developer had decided, quite rightly, to turn this occasion into a very different Sunday lunchtime treat for their guests and added to the big occasion atmosphere by inviting TV film and outside broadcast crews. In fact, the company responsible for Fred's TV programmes had three crews recording the event and Sky News took it live.

The surrounding streets were jam-packed with members of the public keen to see the 'downing' and no vacant local vantage point could be found even as much as two hours

It'll go in a minute.   Any minute now,   'Did you like that?'
                                          don't miss it.

Roger and Mick Barry each collect a trophy.

Roger Dibnah sports his dad's top hat.

Fred and Eddy get to work drilling holes in the telegraph pole props at Burtonwood.

Preparing the fire at Burtonwood, Fred throws on the diesel while Roger Murray looks on.

Fred and the Burtonwood tower.   The fire is lit.

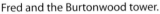

before the climax of the event. The representatives of Greater Manchester Police had a prime view with their flying machine circling the fated structure and several other officers mingled with the expectant, excited and good-natured crowds.

The invited guests and VIPs, in the secure compound at the base of the chimney, had been well and truly fed and watered by the courtesy catering unit and, having been allowed to mingle with the demolition crew and even meet the 'top man', they were slowly moved back behind the safety barriers by 'hi-vis'-jacketed staff, as the clock approached the appointed time of 1pm.

With the appointed time approaching, and the VIP watchers all moved to a safe distance, Fred pronounced that this burn would last for no less than twenty minutes and no more than twenty-five. The match was struck and, within minutes, the blaze took a good hold, aided by the diesel oil and, as if in an act of defiance of its eventual fate, the chimney began to smoke profusely as the timber packed inside the structure ignited.

The sky was black with smoke and the watching public downwind were showered with soot carried in what would be the great chimney's last-gasp breath. The huge crowd of spectators were enthralled and almost completely silent, as the fire roared onward and upward with the black cloud turning to a hotter, lighter colour and the heat created causing the nearby watchers to take a step or two backwards.

Burtonwood going. Note Fred running bottom left of picture!

Burtonwood gone.

Fred had confided that, when the end of the chimney's stand was almost nigh, he expected a brick or two to split and that would, he prophesied, be the signal heralding the collapse of the stack. He had also told the Press and TV cameramen to listen and look very attentively as he would not be able to repeat the show should they fail to capture it on film!

As twenty minutes into the burn approached, the crowd were, it seemed, even quieter, and several began squinting nervously into video and still camera eyepieces. At twenty-two minutes 'in', there was heard a single sharp crack, resembling that of a hunting rifle being discharged, following which the growing excitement among the onlookers became almost palpable.

Within seconds of that tell-tale retort, a huge split opened in the back of the chimney and a section of brickwork fell from the front, in the direction of the intended fall, and in the twinkling of an eye, in fact during the duration of just ten frames of electronically exposed film, it was all over.

The Lion Mill chimney was no more, and it lay in a smoking dusty pile, exactly where Fred had said it would. All that remained was for the demolition man's 'brick-cruncher' machines, which had been parked nearby in anticipation, to grind up the remains of this once-proud structure.

Within days the bricklayers, joiners and other artisans would be starting to put up the new houses, removing forever these particular traces of our industrial past.

As for the crowds, they went on their way happily, having been well-entertained with many of them, including this correspondent, proudly carrying a soot-covered brick as a memento of what would be, for most, a once-in-a-lifetime occasion.

General Patton's revenge

Let us return, for a moment, to the demolition of the aforementioned control tower, which was situated on Burtonwood Airfield, an ex-RAF (and USAF) operational base on the outskirts of Warrington. The flattening of the tower was to make way for the inevitable retail park.

The airbase opened in 1940, just in time to supply Spitfires for the Battle of Britain, and was probably the largest military base in Europe during the second World War. With eighteen miles of surface roadway and a peak of 18,063 personnel, this huge site had a massive impact on the Warrington area and its population.

Many thousands of American service personnel were based at Burtonwood as it was the maintenance and supply base for the USAF in Europe. It repaired and modified equipment, especially aircraft, for use by seventy US bases, reputedly bringing production line methods to Britain for the first time. The very famous General George S Patton was also associated with the area during the second World War and especially in the run-up to D-Day. This job was to have a sting in the tail, or perhaps sting is the wrong word!

Fred was contracted to demolish a very sturdy concrete control tower, it being one of the last vestiges remaining from those far-off days, but it was now standing in the way of the developers. Nothing at all went wrong with the 'burn down' as that superb set of black and white photographs, taken at the time and reproduced here, clearly show. In fact it was a textbook operation, except that flying from the famous base had not, it seemed, been completely discontinued.

Unbeknown to Fred, his crew, the watching dignitaries and the public who turned out in great numbers, a family, or should we say a flight, of ducks was happily living in the top of the old tower. The rumble caused as one of the tower's legs collapsed, preceding its total disintegration, seemed to act as a wake-up call – or should I say 'order to scramble' – and the startled ducks took flight. They embarked upon a bombing run, on a flightpath that took them directly over the watching crowd, the incident proving without doubt that sudden fright can be a stimulus to ducks' bodily functions.

5am knock-up

Dropping chimneys is, of course, dangerous work, and therefore the unwarranted interference of other people is highly undesirable. Neil Carney recalls an occasion in Cockermouth, Cumbria, which could have had a disastrous outcome following one such unauthorised act of interference:

In 1993 we were contracted to demolish a chimney just outside Cockermouth and, after several days on the usual preparation work, we had returned late to our lodgings on the night prior to the event and retired, only to be woken at 5am by our landlady.

She informed us, rather abruptly, that the police were at the door and wanted to speak to Fred, as there was a fire at the site, near to the base of the chimney. We dressed hurriedly and raced off to the job without even having time to grab a brew. On arrival we saw a huge fire blazing at the bottom of the chimney and, some way off to the side of the chimney, the flashing blue light of a Cumbrian Fire Service appliance.

The fire brigade, not being able to ascertain in the pitch darkness the direction in which we intended to drop the chimney, were quite rightly having none of it and had, in the interest of safety, decided not to move in to extinguish the fire until they had the chance to talk to Fred. The emergency services were of the opinion that some enterprising individuals, probably well-fuelled with alcohol on the Saturday night, had decided to do the job for us.

We had not built the necessary bonfire to burn the props, but the timber to do so was on site and these DIY demolitionists had dragged some of that material to the side of the chimney and lit it. By the time the police and fire brigade got there the perpetrators of the stupid act were nowhere to be seen. They had done a runner, probably not realising that they had put the lives of the firemen and police at risk, not to mention us guys who now had to sort out the problem.

Fortunately, the fire they started was nowhere near big enough, but it had removed and considerably weakened a lot of the supporting props. Now one very important thing about this method is that you don't remove the last few bricks on each side of the opening until the 'drop' day. In this case that golden rule saved a disaster.

The chimney remained standing, and a few extra props were hurriedly, but very carefully, fitted into place. The interference of those idiots had caused a big delay, as further firewood had to be procured and brought to the site. In fact, we struggled to find enough and really should have had a lot more.

The drop was delayed and the VIP guests stood around chatting while we prepared a new fire. They were getting more and more fed up and had started to drift away. The new fire was eventually laid and it burned completely away, roaring well up the stack but without causing the chimney even to seriously crack, never mind drop. The fire just wasn't big enough.

There was nothing else to do but to cut out a little more brickwork and, as we did so, we finally got some well-deserved good luck. This chimney was standing on a square plinth, and that portion was cement-rendered so, as we carried out the dodgy manoeuvre of removing bricks on either side of the opening, to what was now an unpropped chimney, tell-tale cracks started appearing in the cement coating.

Fred's last chimney topple – Lion Mill, Royton, Manchester – 9 May 2004.

As the cracks got bigger and more frequent it was obvious that our chimney was finally on the move. About two minutes later, we had moved away to a safe distance, and a huge tell-tale crack appeared in the chimney proper. Then down it went, falling exactly where we wanted it to. Unfortunately, it was almost dark by this time, as a consequence of which, most of the spectators had gone home!

'Concrete Bob' has the last laugh
The great civil engineer Robert McAlpine was a pioneer in the use of concrete in building. The beautiful Glenfinnan Viaduct in Scotland was the first such structure to be built in the

UK using the material. This is an impressive curved creation, 100ft high and 416 yards long, consisting of twenty-one arched spans of 50ft each and carrying a section of the picturesque railway route from Fort William to Mallaig. It is a line that in the summer months is still used by steam trains.

The gentleman was more often referred to, not by his full title of Sir Robert McAlpine, but by his nickname. This fact was for years celebrated by the nameplates carried on a Class 37 diesel locomotive that displayed 'Concrete Bob' on one side and his full title on the other. His techniques became widely used in the construction industry and helped to establish reinforced concrete as a universally-accepted building medium.

The material had been used in the late 1960s to construct a 300ft chimney on Canvey Island in Essex, which utilised a massive 2500 tons of the stuff. This smokestack had become redundant, so Fred was called to see if his method would be suitable for downing the monster. The prognosis was good and, apart from taking special care of the reinforcing bars within the concrete, Fred thought that the job would be nothing out of the ordinary.

The site was to be cleared to make way for the construction of a new Safeway supermarket and the Canvey Island demolition team consisted of Fred with fellow-steeplejack Simon Warner and his long-standing assistant in such matters, Mick Berry. Great excitement and anticipation, as always on these contracts, greeted the arrival of Fred's team. The boys arrived in Essex on 2 September 1997.

Fred established just where he was intending to 'drop' the chimney and then agreed the relevant safety precautions. The construction boss then issued instructions to his team. They were, he told them, to get new crushed white stone and build a roadway and car park at a suitable distance for the use of the VIP visitors, who were all invited to view the 'drop'.

The Canvey Island chimney. Note the A-frame used to support the jackhammer.

Fred chisels away in order to fix another prop on the concrete chimney while Mick Barry looks on.

In addition, a scaffolding grandstand was built and a tower constructed for the use of the local TV news crew. 'In fact,' recalled Simon Warner, 'nothing was left to chance.' He even remembered that the firm had booked the fire brigade to be on hand at the appointed time to damp down the dust by creating a curtain of water on two sides of the chimney, and to achieve that they had laid out hoses with stop ends and full of vertically pointing holes. It was a very rare late summer, a very hot and dry one and, as we started work, the September weather was good!

The preparation was, Fred figured, likely to take fourteen days and that included estimating for him being off-site to attend an appointment with the magistrates in Bolton over the matter of him burning wood in a smoke-controlled zone. There was to be a steam traction engine rally in Essex, which coincided with the estimated finish date of the work, and Fred intended taking the roller there while Mick and Simon drove back to Bolton. It was to be the normal cut-and-prop technique except that Fred wanted to drill and cut the vertical reinforcing steel rods around the structure as well. The thought was that, having cut some, the others (around the back, away from the intended fire) could be severed as the fire got going.

They were all duly marked up and the boys got down to the task of cutting the chosen ones out and then chopping out the sections of concrete on the side of the fall. They would then put in the props, which would temporarily support the chimney. There was on site a great pile of scrap timber, which had been brought by the contractors for use in the burn. The local TV people turned up to do a little 'before' filming and the Fred dutifully explained to them the intricacies of the project.

Fred went off to Bolton and duly returned on 17 September. He confirmed with the boys that the drop was still set for the next day and, while doing so, he observed that the

Missed it! The Land-Rover is safe and sound.

A day early but down nonetheless! Note the wood for the fire, now surplus to requirements. From left: Mick, Fred and Simon.

contractor's bunting was no more than fluttering in the moderate breeze. 'Good,' he thought. The temporary car park was looking like it had been there for ever and the contractor guys left on site were all happily in the cabin doing what many think builders do best, having a brew.

The boys started to show Fred what was left to do and each went about his allotted task, doing a bit more jack hammering at the extremes of the cut-out section and checking the props. It was just after that, Simon recalled, that it started to rain – not water but little tiny bits of concrete!

The boys, who were all working on the same side, stopped work for a minute and nothing happened, so they decided it was just a bit of loose from the drilling. They started to drill again and then the concrete rain started in earnest, but this time they were not tiny specks but quite big lumps, and they were getting bigger by the split-second and a lot more numerous.

Fred yelled: 'Run! We're not in charge of this bugger!'

The boys started by instinct to run in the direction they knew to be the opposite one from the intended fall. Mick, according to Simon, broke the four-minute mile and became the first man to do so while wearing a hard hat and steel toe-capped shoes. Panic! At about 60ft from the crumbling chimney, Fred tripped on a bit of reinforcing wire and was down on the deck and 60ft behind Simon. Mick, he says, was already in the winner's enclosure.

Simon remarked that he had often heard the expression 'frozen to the spot' and at that moment said he was. His head said 'run back and help Fred get up' but his brain didn't tell his limbs to move and, for an instant which seemed like an age, the pair just stayed where they were. Fred slowly rolled over and then struggled to his feet and still they both just stared as the chimney then dropped vertically down with a mighty whoosh, all 2500 tons of it – right before their eyes. An almost out-of-breath Fred just pointed and shouted above the noise: 'Bloody hell, a day early but right on target!'

As if the shock of the premature drop was not enough, Simon then recalled that he had parked his Land-Rover only yards from the base of the chimney. How, he thought in a flash, do you explain that to your insurance firm? If the sudden collapse was a surprise to the three demolitionists it was an absolute bolt of lightning to the boys in the site office.

As the chimney fell before his eyes the foreman was on the phone to the boss assuring him that Fred was back on site and that all was well for the big event. What he is reported to have said next as he saw the crash we will leave to your imagination, as it was definitely not printable.

'What happened next,' said Simon, 'was like something out of The Wacky Races.'

Four of the guys from the site office rushed outside and jumped on anything with a motor, dumpers, JCBs etc then all raced towards them through the huge dust cloud to discover if they were all still alive.

Remember the water curtain to spray down the dust was, like the demolition, booked for the next day. Simon's Land-Rover? Well that escaped, but only just, and he recalls they did get paid. Fred maintained they should have had a bonus for being a day early, adding: 'You can't trust concrete. Give me good old-fashioned bricks every time'.

He went off to the rally with the roller and the boys returned home with Simon still thinking about what, in different circumstances, he would have said to the insurance agent!

Fred shares his favourite magazine with engineers at the Midland Railway Centre near Ripley, Derbyshire.

Fred with a steam locomotive crew at the Great Central Railway, Loughborough, Leicestershire.

Handsome young devil, Fred Dibnah age twenty-one.

Demolition Man

In and around Bolton there has for years been no confusion as to who deserves that title, and it is not a certain American actor with a propensity to mumble.

That guy may have been the star of a film called *Demolition Man* but, in reality, the only man who should carry the title, according to folk around Bolton, is Fred's great friend, Harry Forshaw. Listen to enough conversations about Mr Dibnah's chimney-toppling feats and it will not be long before you also hear the phrase 'yon demolition man'.

The genteel and quietly spoken Harry Forshaw is the man they are referring to and, with his family firm, he was involved in the removal of tall objects Dibnah-style right from the beginning.

They toppled their first chimney together at a local mill in 1971. Harry is not sure exactly how many structures he has employed Fred to topple but he thinks that it is probably getting on for twenty. That number also includes towers as opposed to chimneys and, on all counts, Harry makes an interesting point often overlooked by those who came to watch those gladiatorial spectacles.

He said: 'Although Fred had finished and all the dust had died down, that was nowhere near the end of the job. We then set to work moving all the waste, and that would always amount to shifting hundreds of tons of the stuff.'

Below ground, Harry Forshaw deep under Lancashire with Fred.

After the drop, a Rochdale Mill chimney is no more! L to R (Starting with guy hands on hips) From the left, David Lunn (with hands on hips) Harry Forshaw holding his son Christian, Lorna Dibnah, Lyndon Forshaw, Jonathan Forshaw and Fred.

All that was left after a chimney job at Farnworth in 1981.

'Fred always said that he was sad at knocking down things which the builders of long ago had toiled over and, of course, that was true, but in a way it was not always the absolute end.'

He continued: 'I used to often tell Fred what we were doing with the rubble and, in a lot of cases, I was able to give him some consoling news in that a lot of the waste went back under new buildings and roads, so I suggested that it was not an end but a beginning!'

'"Aye, you are right, Harry, but try telling that to the poor old buggers who toiled for next to nowt to build the thing," was his usual style of answer.'

Harry recalled Fred had his own special logic and you could not really argue with it.

The two men were not just work colleagues, they were friends and as such shared similar interests, so much so that the families went on several visits together to the likes of Ironbridge Gorge and Beamish Museum.

Another interest they shared was the history and even, on occasions, the underground exploration of the now-defunct but once great Lancashire coal fields. In line with those interests, many of the articles in Fred's comprehensive collection of artefacts were found by Harry during the course of his everyday work. He had rescued them and thus saved them for posterity.

Not every job he carried out with Fred went exactly according to plan. In particular, Harry highlighted the problems associated with two types of towers or chimneys that always presented a few problems.

One was concerning things made with reinforced concrete and the other problem was a structure with a square base. Concrete was a building medium that Fred mistrusted every time he got to grips with it. In addition to the famous concrete chimney in Canvey Island which nearly wrote off Fred, Simon Warner, Mick Berry and the Land-Rover when it fell prematurely, Fred had other close calls with the stuff. He did get rid of a big concrete chimney in Trafford Park for Harry's firm but his employer recalls that all the time Fred was on edge and kept saying: 'I hate working on this bloody stuff'.

One near-disaster Harry recalls was at Laburnham Mill in Bolton, although it did not involve concrete. It concerned a big, square tower with a huge water tank built in the top of it. The gang was also to demolish a chimney on the site in the time-honoured way, but the tower was to be felled first. Fred had spent the usual amount of time preparing the thing – about seven days, said Harry – and all was set.

The crowd gathered and, when all was safe, as usual Fred lit the fire. 'It roared away good style,' Harry recalled, and anticipation grew among the watchers. But after a while the fire began to die and the tower still remained standing. It became apparent that all was not well.

Harry said Fred was very concerned and 'well put out'. Something had to be done and quickly as, even though the tower still stood, it was in no condition to be left as it was. The crowd began to disperse and, when people were all out of the way, Fred decided to get to work on enlarging the 'slot' with his jack hammer.

Harry said they agreed that he should watch and raise the alarm as any possible cracking could not be heard by Fred over the sound of the air hammer, the compressor and something he never then admitted to, the onset of partial deafness.

Fred was well into his rhythm and had removed a fair bit more when Harry said that he saw the slightest of movements followed by a trickle of dust.

The mines exploration team, left to right: Dave Turner, Mark Wright, Alan Davies, Harry Forshaw and Fred.

Below: Ready for the off, Fred prepares to light the fire under a chimney at Ocean Mill Great Lever in 1982.

Quick as a flash he dashed forward and, as he did so, shouted for someone to knock off the air. Just in time he got a hand on Fred's collar and pulled him away. The pair then turned and ran, literally for their lives. Alarmingly, as he remembered it, the whole thing then collapsed just like it should have done in the first place.

That mill also had a bit of a sting in the tail when it came to the chimney, said Harry. He told of the way the chimney came down and it, too, was not as planned.

Fred had spent six days getting the chimney prepared and all was set for 'the drop' before a big crowd on the Sunday morning, finishing in time for a pint!

At about 5am Harry said he got a phone call from the police to go to the mill site. On arrival he saw that the stack was down with the rubble lying exactly in the place where it should have been!

The fire brigade was in attendance and the fire crew boss gave Harry an explanation. By all accounts someone the worse for drink had seen the prepared chimney presumably as he (or they) were on the way home. The foolish people had lit the fire but, instead of leaving it at that, had also thrown into the flames a large liquid propane gas container.

When the fire brigade arrived they saw the gas container in the flames and prudently evacuated the area, fearing an explosion. In the time it took to do that, the supports burnt through and the law of gravity did the rest!

Harry can also throw some light on another thing which will for ever be associated with our Fred, that being the question saying: 'Do you like that?' often quoted as being: 'Did you like that?'

While Harry says that no doubt both are correctly attributed to Fred, the original use was as follows: they were working on a chimney at Era Mill in Woodbine Street, Rochdale, and it was another Sunday job. As the fire got to its height, the 200ft chimney started to topple and, as all the watchers looked on with amazement, it seemed to hang suspended in mid-air.

It was at that instant, and before the thing finally succumbed and fell asunder, that Fred turned to Harry and said for the first time the immortal words as he pointed at the stricken stack: 'Do you like that?'

When asked about that modern curse 'cost effectiveness', Harry thought he would have to say hand on heart that blowing a chimney with explosives was the cheaper option.

But he countered: 'When Fred did it his way, what was just another job became a truly great occasion loved by many.'

He added that he was happy to have been involved with Fred as both a workmate and a friend, and that knowing Fred well was the greatest experience and a real pleasure.

Fred chops out a bit more while Neil Carney keeps the standing area safe.

Up, Up and Away!

Heights were never a problem to Fred Dibnah, and his climbing exploits are legendary and numerous – but his first hot air balloon experience was, in his own words, 'something completely different'.

He was invited to take to the air by his long-standing friend and fellow raconteur, Fred Fielder, a gentleman who has been one of the north-west's leading radio presenters for more years than he would care to admit.

He is also a very widely travelled man and as such it comes as no surprise to learn that he is a Fellow of the Royal Geographical Society.

The flight was a very private affair and it was kept low key for a couple of reasons, not least of which being that Fred (Dibnah, that is) was in the middle of filming a TV series. A spill on landing and any broken bones would not have gone down well with the film production company.

He was, however, in very good hands because Fred Fielder was at that time a regular hot air balloon pilot and an accomplished balloonist, having even flown in India with that other intrepid adventurer, Roger Murray – but that is another story!

Young Mr Fielder first met Fred D at a steam event in the 1970s and the two men 'sort of kept in touch', he recalled, but they really became acquainted a year or so later when Fred F was invited to a function at which Fred D was guest speaker.

The MC declares a draw as Fred F carries our Fred to the beer tent.

Fred Fielder's black balloon, with the skull and crossbones, in which Fred D had a flight.

The two men got on famously and, after dinner drinking sessions around a convivial bar being what they can be with the right participants, the pair talked on well into the early hours. Young Mr Fielder was instantly drawn into the company surrounding the steeplejack.

That evening was long before the advent of the breathalyser. Fred F recalls that after the long-suffering bar manager had announced that he was going to bed, he was the one who had had the least to drink and therefore finished up driving Fred D home.

He does not recall much of what happened but he says it was the first time in his life that he drank white wine and Guinness! As a consequence, his car stayed at Radcliffe Road and he walked home wobbly, listening to the dawn chorus.

During the course of their late night or, more correctly, early morning conversation, Fred D was fascinated to learn that his new friend was at that time a professional hot air balloonist and airship pilot. In fact Fred Fielder has flown over the Amazon, the North Pole, Africa and most of the India sub-continent as well.

The balloonist's conversation kept Fred enthralled and all of a sudden he asked: 'Do you fly around this part of the world?'

When Fred F said he did, our hero asked: 'Is it a big black one with a skull and crossbones on it? I've bloody seen you sailing low over this place!'

Then Fred D remembered an incident from a year before.

'Did you shout to us when we were on top of a chimney in Farnworth?'

Fred Fielder in India.

'Yes I did,' said Fred F, adding: 'There were three of you sat smoking on the top of a chimney. You asked us where we were going and I gave the classic balloonists' answer that I didn't know. I said you should ask us where we'd come from, and you did.'

But the answer was not what Fred D expected.

As there were three in his gang and three in the basket, it seemed only right to Fred Fielder to give the steeplejacks the flip answer: 'From a stable in Bethlehem!'

The balloonist recalled that Mr Dibnah replied in a suitable manner, using words of Anglo Saxon origin! There was in truth a very vague biblical connection as the balloon had taken off that day from the Moses Gate area of Bolton.

Fred Fielder continues the story.

'Naturally Fred Dibnah was inquisitive about ballooning and it was inevitable that he wanted to have a go. However prior to arranging that we staged an ad hoc event called "The Challenge of the Freds", between his steam roller and my balloon.

'We were in the beer tent (there's a surprise) chatting over old times at a small show in Urmston when we hit on the idea of a tug of war between his roller and the balloon. We clued up the announcer to wind up the crowd and off we went.

'The balloon was already tethered on the blind side of the crowd and so although airborne, it was going nowhere.

'We tied another rope from the other side of the basket – the side in view by the audience – and I let it drop down so that Fred could then make a great show of lashing it to the back of the roller.

'That act was accompanied by loads of air punching with his fist and plenty of "We'll bloody have him – I'll show him who's the boss" etc. The crowd loved it and Fred managed to get one of the roller's back wheels to spin as if it was skidding, with lots of steam and hoots on the whistle going on at the same time.

'Meanwhile, the balloon's burners were being turned on and off creating a great roar while, at the same time, we kept the rope tight. It was a good ten-minute special event and those who saw it will I am sure never forget it. What these modern health and safety bods would make of it I don't know! They probably would have locked us both up and thrown away the key!

'The master of ceremonies, who incidentally did a great job working the crowd, eventually called a halt to proceedings, declaring the contest as being drawn.

'We again retired to the beer tent – in fact, I picked Fred up and carried him there!

'We did some time later make a very enjoyable short flight and Fred absolutely loved every minute of it. He was like a lot of people – a little reticent as we took off and seemed to be holding tight to the basket side, but after a short while he was as relaxed as any seasoned balloonist and perfectly at home with the height.

'He did, however, comment that he was happier sitting on the top of a chimney because it was fixed to the ground. I pulled out all the stops and made sure that the landing was not too rough. He said that he wanted to do it again some time but, sadly, we never did get the chance.'

Another thing the two Freds had in common was engaging in the entertaining art of conversation.

Fred F pointedly remarked: 'The trouble today is that if you ask a lot of strangers who you have just first met more than three questions, they think you are from the police or

Fred with his preferred style of
'titfer'!

Fred Fielder, fellow-raconteur, and Fred D were good buddies.

DSS, and clam up. That's a shame as conversation can be a very entertaining and rewarding pastime.

'One thing Fred and I used to like to talk about was political correctness which neither of us approved of greatly, and also we loved the art of giving people good nicknames. I would phone him if I heard a new good one and he would often respond by telling me one that he had learned.

'One of his favourites was, in fact, one of my dad's old ones and it originated, like so many funny sayings, from Liverpool Docks.

'There was a guy in one of the work gangs who they called "Sammy the Crab". The explanation why illustrates the fabulous humour of those guys. It seems the man would never work overtime and when asked to he would always give the same answer: 'Sorry, can't stop late, one of my nippers is bad!'

There were literally hundreds, like "Tab Hunter" for the guy who was always bumming ciggies and then there was the docker they called "The Destroyer", because he was always chasing after a sub!

Fred's nickname for the Army Catering Corps was "Andy Caps Commandos" and the people who worked the sands at Blackpool giving kids donkey rides were collectively known as the "Donkey Wallopers"

Having a conversation with Fred Dibnah was always a pleasure and more often than not, it would have an amusing twist to it. He was truly the font of all knowledge and although he never thought so himself, he was a very funny guy.'

Fred Dibnah, or his actions, could also be conversation stoppers.

Fred Fielder recalled this gem.

'I invited Fred and his wife to the celebration of what I should describe as a significant birthday of mine. He was due to arrive about mid-afternoon. The party was going well and the weather was kind so some people were imbibing outside in the Lancashire fresh air while others were sitting in groups inside, chatting away merrily.

'Fred arrived and he was well scrubbed up and it seemed wearing a snazzy (but definitely not him) new black and white check flat cap.

It seemed the 'titfer' had been bought especially for this occasion by his lady wife. Fred was seen to emerge from the car a few paces behind her, wearing that incongruous vestment. Mrs Dibnah went onto the patio at the rear of the house where she had spotted some friends while Fred settled, a pint of the black stuff in hand, in the front room with a group of guys who were all anxious to ask him about his latest bit of telly!

'After an hour or so, Mrs Dibnah wandered into the house to see if Fred was OK and consequently spied him in the front room. She was just starting to wonder what was different about him when the little boy from next door rushed through the front door. He excitedly declared: "Look what I found in the field at the side", while waving a black and white checked cap about.

'He then added: "someone must have chucked it over the hedge an' it's a brand new 'un".

'All eyes turned to Fred, who was grinning a little embarrassingly and sat there wearing one of his trademark old flat caps. The look on Mrs Dibnah's face told its own story. See what I mean about a conversation stopper. God bless him!'

Fred with Mike Bossons outside Metropolitan-Vickers' Manchester works. Note young Jack Dibnah on the footplate.

'Atlas' – Fred's Favourite Engine
Jim Stevenson, Film Maker and Writer

From his very first encounter with traction engines, Fred Dibnah always looked for perfection, and the superb condition of both of his own restored machines bears adequate witness to that belief. However, anyone who knew Fred will tell you that after his first encounter with 'Atlas' he was smitten, and it was indeed love at first sight, with the big Fowler B6 'Super Lion', works number 17105, registered VM 2110.

Fred had seen 'Atlas' at several events and he was forever talking about it. No one would ever have described Fred as a man even remotely obsessed with material possessions but, having encountered 'Atlas', Jim Stevenson, another of Fred's friends who is also a writer and sole proprietor of the film company Moonlight Productions, takes up the story:

I first met Fred in 1974 during a visit to an evening steaming event at the rear of the Red Lion public house in Ellenbrook, Manchester. Neither of us had brought along an engine so we kind of drifted into one another's company and accordingly strolled around looking at the vehicles on show. As many others have intimated, it was not difficult to become

Fowler road loco 'Atlas' recreates the type of load it would have pulled in its working days, in this case a transformer.

The engine outside Metropolitan-Vickers' Manchester works.

friendly with Fred – in fact, quite the opposite was true and I left that evening having promised to keep in touch with him.

During the course of some subsequent refurbishment work to Fred's steam roller, I regularly visited Bolton and I recall, in particular, that I assisted with the refacing of the slide valves and helped with the 'grinding in' of the regulator valve. Some years later I visited the Isle of Anglesey Traction Engine Rally and Fred subsequently told me that he was impressed by the fact that I had 'single-handedly' road-driven my steam roller to the event, a round-trip of over 200 miles. In fact I did that for two years on the run in 1979-80 and my first Anglesey journey became the subject of a book entitled *A Head Full of Steam*. My machine back then was a 1924 Aveling & Porter E-type road roller 'Cinderella', which I restored between 1975 and 1977.

'Atlas' is seen setting off from Fred's house in Bolton.

But back to Fred's involvement with 'Atlas'. When, out of the blue, the engine's owner, James Hervey Bathhurst of Eastnor Castle, asked him to carry out some repairs on it, he must have thought his dreams had come true! The Fowler B6 was therefore delivered to his workshops. Following the work he got to drive 'Atlas' during the making of a film depicting its 'trial' return journey to the Cheshire Steam Fair in 1992 entitled *Travels With an Old World Atlas*, and on that run he was joined on the footplate by Mike Bossons. During the run we made a planned diversion in order to take the loco from Bolton to Trafford Park, Manchester, to visit the premises of engineers Metropolitan Vickers Ltd, who used it to haul heavy loads (often transformers) around the country during the late 1920s.

The big road locomotive featured in a further Moonlight Productions film when it travelled 'light' from its base at Eastnor Castle in Herefordshire to Gloucester and then hauled a 40-ton road train in the reverse direction, over the Malvern Hills. This was in support of a charity event organised by a local radio station.

Fred joined the ensemble for the loaded 'road train' trip when he again piloted the 'Big Un', and that epic 1993 journey was also captured on film, entitled *The Big Road Atlas*. Moonlight Productions has also produced a film in order to commemorate the National Traction Engine Trust's 2004 Golden Jubilee Event, entitled *Going For Gold*. Jim Stevenson's films are still available from Moonlight Productions and he can be contacted on 0161 969 7783.

Derek Rayner is the Technical Editor at *Old Glory* magazine and we are indebted to him for compiling a fact file on the Fowler B6 road loco and some additional information on Fred's convertible.

'Atlas' - a Fowler B6 Super Lion road loco – was new to Norman E Box Ltd, a haulage contractor who had offices in both Manchester and Birmingham. The engine was despatched to that firm from John Fowler & Company's Steam Plough Works in Leeds during May 1928. The engine was fitted with a crane jib attachment (for use with a removable jib) and carried the necessary rope drums etc from new. The lugs for this are still retained on the front of the smokebox.

The B6 spent the next decade moving loads of enormous weight and physical size such as transformers, accumulators, boilers and condensers the length and breadth of the country. This it did not only as a solo machine but also occasionally in conjunction with other engines. The engine later passed into Pickford's Ltd ownership for similar use when Norman E Box Ltd amalgamated with that well-known company in 1930, and eventually, it entered into preservation.

The major part of the engine's initial restoration was carried out by the then-owner, George Trubshaw of Stoke-on-Trent, who carried out the exacting work mostly without outside help. The engine was first 'unveiled' to an admiring public in early July 1984 at the Elvaston Castle rally, near Derby.

Fred's newly-unveiled Aveling Convertible Traction Engine, Aveling & Porter works number No 7838, was one of four seemingly identical machines supplied to West Sussex County Council. These were 7836, 7837, 7838 and 7839 respectively. They were of a type described in the Aveling & Porter build records as KND 4hp 'motor tractor & roller combined'.

No 7838 was delivered on 4 December 1912 and was fitted from new with a two-tine scarifier No 1256. This quartet were all thus fitted, and they were dispatched from the works with both a set of rolls for rolling purposes and straked wheels for use in tractor mode.

No 7838 has additional entries in the company's records. It is noted that it was Devon County Council, Fleet No 20 of 29 March 1927. It is also noted that the angle segment (on the front of the smokebox) was dispensed with and a cast iron 'combined' type chimney base with flange for bolting on a saddle (number KND 226) and a new chimney complete were supplied by Aveling to the council. At that stage it was thus definitely in use as a steam roller and there is no doubt about it. Hence it is a convertible engine and therefore it has now been restored in its tractor form.

Meet Spiny Norman
also known as the Sputnik

There were occasions when Fred was unable to take down a chimney using his tried and trusted method of fire and gravity, mainly in locations where there was no room to 'drop' the structure either safely or without fouling something like a river or canal.

The alternative was to take the chimney down brick by brick – very labour-intensive work. In itself, it was a very time-consuming operation, but moving the scaffolding as you lowered the structure was even more of a pain in Fred's view. And so he put his active mind to work and Spiny Norman was born, although Fred and his team can take no credit for the name, which they nicked from a character in the *Monty Python* TV series.

The idea was first tried by way of a scale model constructed in the yard at Bolton at a time when the Russians were launching the 'Sputniks' into space. On seeing the model, Spiny Norman thus acquired an 'also known as' name.

Fred's clever idea – reputedly first drawn on the back of a cigarette packet – was intended to speed up the time taken to move the scaffolding, making the job easier and so, hopefully, giving Fred a better chance of seeing a profit at the end.

Simply speaking, the apparatus was designed to carry the work platform scaffolding on a strong rigid frame, which could then be lowered down the chimney as the demolition progressed, thus alleviating the need to rerig the scaffold every time a few more courses of bricks had been removed.

It worked a treat and was another great example of Fred's ingenuity.

Fred loved to put his ideas on paper before putting them into practice. He is seen at the drawing table in his living van.

The mock-up of Spiny Norman in Fred's yard. It is easy to see where the name Sputnik came from.

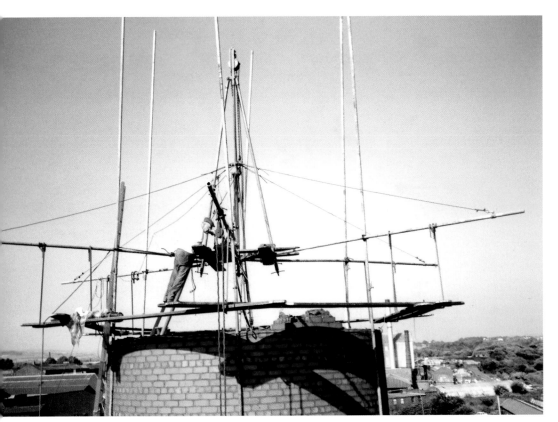

A job well done, thanks to Spiny Norman, as the demolition is all but finished.

A Musical Interlude

David Devine

Ask David Devine to tell you what a 'Wigan kebab' is and you will get what is to him a very obvious answer: 'Three meat pies on a stick'. That's because, although he is a Lancashire man through and through, he is firstly a proud Wigantonian. He is the leading light behind the popular website Wigan World (www.wiganworld.co.uk). That modern news organ has, since its inception, championed the causes of Fred Dibnah and his many acolytes. David Devine can fairly be referred to as a Fred Dibnah archivist and, in keeping with that title, has a barrowload of stories to tell, in addition to a fine collection of photographs. David writes:

It has been my pleasure to have known Fred Dibnah since 1969, and I've been fortunate to accompany him to many events. I have also been mixed up in several escapades, a great many of which involved my late brother, Ken, who Fred often described as a 'frustrated steeplejack!' Ken was, in fact, a virtuoso musician and, in his latter years, a violinist with Manchester's famous Hallé Orchestra. But, if the slightest opportunity presented itself to him, he would go climbing with Fred and, like him, my brother appeared to have no fear of heights.

As we would leave the house to meet up with Fred at a worksite, Ann, my sister-in-law, would elicit from both of us a promise not to do anything dangerous and to keep off chimneys. For years Ken, in particular, broke that promise regularly and got away with it – that is, until the occasion when his photographic skills gave him away.

Having collected a particularly good set of prints from the processor, he got a little carried away and decided to show them to Ann. She studied them and pronounced them as being very impressive, and then her face darkened a little as she noticed that something familiar to her appeared in a shot taken 'looking down the chimney' towards where Fred was working. 'Ken Devine!' she exclaimed, 'You have been up that chimney or someone else was wearing your shoes!' There, clearly in shot, were Ken's lower legs and shoes. His wife had him 'bang to rights'.

Virtually everyone who ever worked with Fred has a chimney demolition 'funny' and mine, as you would expect, is not a tale from a perfect job. In fact, this particular chimney was in a very restricted situation. Fred had weighed up the job with his usual diligence and we could not have been aware of the problem that awaited us. As the chimney started to fall (in the right direction) one large lump broke away from the top and headed down on its own. It struck and then went through the roof of a brick-built shed next to the job. No problem, we thought, as the shed was out of use. This part of the factory was derelict.

Wrong! The shed was still being used as a bicycle store and Fred's insurers had to sort out a claim for the damaged bike belonging to a very irate storeman and, in addition, a claim for damage to the roof. Now that was a little out of order as several days later Fred received an enquiry from the firm asking for his best price to knock down that very shed!

The Rev Barry Newth recounted several amusing tales at the 16 November 2004 funeral of our dear departed friend, one of which concerned him and a colleague turning up as promised at a church fete on the steam roller, but they were very late and a little the worse for drink. He kindly did not name Fred's accomplice at that time but, as the saying goes, confession is good for the soul. It was me and, boy, had we had a great session!

One of David Devine's favourite pictures – Fred posing by his Land-Rover in the mid-1980s.

Fred on the top stage of five, 1980.

In his own way, Fred was very fond of music, so he would chide my brother constantly on the subject of varying musical tastes. On one particular occasion we had been returning to base with a roller, having successfully completed some project or other, the details of which escape me, except to say that we were hungry, thirsty and covered in black soot. That roller used to have a mind of its own and, amazingly, it had a habit of stopping outside the Lever Bridge Inn.

Now we naturally had a few drinks and, while doing so, transferred our dirty fingerprints to a few sandwiches, a couple of pies and the odd bag of crisps. Everything in the world was well and we were all enjoying the conversation and the ambience that was, back then, typical of the traditional British pub when Fred suggested: 'What we need now is a sing-song. Come on, Ken, you're the musician. Get on that piano in yon room.'

Fred at work, 1981.

At that point we all picked up our pints and headed for that place especially beloved by Ena Sharples and company in the early episodes of Coronation Street, the uniquely British 'snug'.

Ken complained that he had no music so could not play, which cut no ice with anyone. 'You're a pro' and 'Get on with it' were the typical remarks that I dare repeat. Now, you never know who is in a pub listening to your conversation and, in truth, as the drink goes in you care even less. There was a very smartly dressed elderly man propping up the bar and, on account of the winter weather, he was wearing a long, dark grey overcoat.

He just motioned to get Ken's attention and then, as the film makers would have us believe the archetypal 'wide boy' or Arthur Daley-type would do, he slowly opened his coat to reveal what looked like an illicit sheaf of papers in his inside pocket. They were, it turned out, a collection of sheet music appertaining to popular pub tunes and even included a couple of hits of the day. That was it! The sing-song lasted all afternoon and a great time was had by all. However, Ken needed to be at the Hallé for a concert that night so we eventually dragged him off the piano and headed for Fred's house, where we had left the car.

We were, to say the least, a little late and, to make matters worse, Ken needed to wash and change into his 'zoot suit' for the performance. Now, that gear was in the back of his car so he grabbed it and, having washed quickly, he tugged on his posh gear and headed for the car. We just about had time to make it to Manchester, but Ken's starter motor thought otherwise, and the thing would not start.

Fred came to the rescue: 'We can go in the Land-Rover.' Which was fine with a by-now very anxious Ken until he remembered: 'Neither of you can drive in that state.' It then transpired that Ken would drive, as he claimed to be sober, and that Fred would sit with him and explain the foibles of the gearbox to him as they went along. I climbed in the back and off we went leaving instructions for the garage to collect Ken's car.

The plan was that we would have a good feed and sober up while Ken scraped the strings for an hour or two and then we would all travel home. But, as we got to the stage door, my brother asked why we did not want to see (hear) him play with the Hallé. Well, that was that! Into the Free Trade Hall we went – our kid in his show gear and us two in our scruff, but we did have time to dive into the gents and have a cursory clean-up just before the concert started.

Now that is not entirely the end of this musical story as, more than two years later, we were in another pub, this time in nearby Darwen, when Fred spotted the guy with the overcoat.

He immediately recognised our group and he again slowly opened his coat while tapping his bundle of music, which was still in that inside pocket. Ken at that point suggested that we find another watering-hole and Fred and I fell about laughing.

Words cannot express how deeply I feel the loss of Fred but, thankfully, like a good many more, I have a host of memories to console me.

Cat and Fiddle

Fred outside the famous Cat and Fiddle Inn, which, at 1690 feet above sea level, is believed to be the highest licensed house in England.

When steam traction engines and steam rollers begin descending long, steep hills there are always a few precautions to bear in mind. Fred's Aveling & Porter convertible traction engine is pictured on the Llanberis Pass in North Wales.

'Nut Roast' and the Red Rose Steam Society

On the edge of Chat Moss in Lancashire, in an area once full of collieries, lies the picturesque village of Astley Green. In the heart of the village stands Astley Green Colliery Museum which, but for the foresight of Lancashire County Council and several leading figures within the community, would have suffered the same fate as the other collieries in the area – total demolition. It is a site that Fred was familiar with and had visited regularly. It has to be said that he particularly enjoyed the fare on offer at the adjacent local chippy and the craic in the local hostelry.

Fred takes off the chimney cap so that it can be polished before the big journey gets under way.

Fred deep in `steam talk' with Richard Fairhurst.

'Made it!' The end of the first day's filming.

A puzzled Fred gives the cylinder block the once-over.

Richard is explaining to Fred the story of the 'Nut Roast' name, but quite where he is up to in the story is not obvious from the position of his hands. Our reporter Sue Mills looks on but, perhaps wisely, she failed to relate what the guys were talking about!

`Full service'. The Aveling gets a flue- clean at Astley after its test road run.

The imposing lattice headgear at Astley is virtually all that is left of the once-mighty Lancashire Coal Field.

It was the realisation of the uniqueness of the 3300hp twin tandem compound steam winding engine that brought the demolition of the redundant colliery to a halt. Accordingly the museum, which houses Lancashire's only surviving headgear and engine house, now has listed building status. The preservation site and visitor centre is the headquarters of the Red Rose Steam Society, an active body, of which Fred was a keen supporter.

The museum occupies some fifteen acres of the old colliery site and is virtually all that is left of the once-great Lancashire Coal Field. The low-lying landscape ensures that the museum's 98ft high impressive lattice steel headgear can be seen clearly from the busy East Lancashire Road (A580). A fitting memorial to days now past, the steam winding engine and its headgear complement the museum's many other industrial exhibits, not least of which is the collection of twenty-eight colliery locomotives, the largest collection of its type in the United Kingdom.

Richard Fairhurst explains why the nuts were well and truly roasted!

The Astley ex-colliery site was chosen by Fred and the film crew as a suitable venue to visit at the beginning of his round-Britain tour. Thus the first real test run of the Aveling & Porter convertible terminated there after a successful and spirited run from Bolton. It was then serviced at the site over the weekend, prior to being low-loaded up to the Lake District, where it was required for more filming. It was on the fascinating coalmine site that Richard Fairhurst renewed his acquaintance with Fred:

I have been fascinated with anything steam or mechanical for many years, probably back to when I was five or six years old. My late grandfather, who was a very patient man and fortunately indulged my childish curiosity, encouraged me or, should I say, put up with my endless questions. He was a great fan of Fred's and so, at an early age, I knew all about the local steeplejack who had restored a steam road roller.

I joined the Red Rose Steam Society when I was thirteen, went to my first traction engine rally that year and became well and truly bitten by the preservation bug. From then on I have been going to rallies, and a major part of my life now revolves around steam preservation. Like a good many, I bought the early Fred books on steeplejacking and steaming right through my teenage years. By the time I was fifteen or so I was probably quite an authority on him. I suppose you could compare it to another youngster worshipping a favourite rugby league player or cricketer. I read and inwardly digested every word.

Born in 1971, I do, of course, understand the need for, and the advantages of, modern technology, but I am also able to appreciate the engineering skills of the past and, like Fred, my hero was Isambard Kingdom Brunel. I joined the Lancashire Traction Engine Club (of which I am now, for my sins, membership secretary and website producer) when I was eighteen, and that marked a very memorable incident in my life. It was at one of their meetings that I had my first conversation with Fred; there I was, in the same room as the guy I had read so much about.

Around that time I got to know David Lomas, owner of Aveling eight-ton roller No 10753 (NU 3051), when I was still at university and had just turned twenty-one. I began helping David with his engine at rallies and events on most weekends until about 1998. By that time I really wanted a steamer of my own and I set out in earnest to look for one and succeeded. On a very cold November afternoon in 1998 at Astley Colliery, I proudly first steamed the Fowler roller I had just bought, it was then called 'Jenny'. It was then, and still is, my pride and joy.

That was on the Saturday and I slept well that night, with plans to road the machine around the village the following day firmly lodged in my mind. I was up early and soon had 'enough on the clock' to turn the wheels. As I was about to take to the road, who should just happen to turn up at Astley with some friends but Fred. He regularly popped down to Astley, normally calling at the pub opposite, and then having a wander round the site. He always came chatting with the boys working there and was genuinely interested in their projects.

The machine I now own was ordered new on 17 March 1925 by Aberdeen County Council as a DN1, 10-ton, 5hp, compound tar-spraying roller. It was delivered to them several months later on 28 May 1925. When supplied new as a tar-sprayer, it was fitted with Fowler-Woods tar-spraying gear, a differential, rope drum, fairleads, rim brakes and a full canopy. It was also supplied with two patent gritting machines, and Hecla tar boiler.

The gritters cost an extra £360 each, and the tar-boiler was a further £114-10s. The engine, with a ten per cent discount, cost the authority £1171, that figure being equivalent to about £40,000 in today's money. During its working life this engine, along with four others supplied by Fowler's, stayed in the Aberdeen area until entering preservation in 1968. All five engines survive today, although two of them are now in 'showman's guise'.

Since the engine was sold into preservation it lived in the Yorkshire area, until moving to Cheshire in the early 1990s, and then finally Lancashire to its present home. When I bought the roller it was a runner, so I never had the work of completely rebuilding it, but I have kept it in tip-top condition and that is something Fred often commented on. He also noticed that I had changed the name to 'Nut Roast' and asked me why I had chosen that 'handle'. He roared with laughter when I told him the reason.

Like all kids with a new toy, I never wanted to put it away. I was forever cleaning and polishing the Fowler; every bit of bright metal was polished on every spare occasion, and still is. I was thinking about a choice of name and couldn't really make my mind up, until one day after the cleaning session it came to me, literally in a flash.

I had been cleaning up the paintwork and I used a light solvent on a rag to wipe over the surfaces. I gave the rag another good soaking and, in doing so, I obviously spilled some of the solvent on my boiler suit. Somebody shouted 'brew up' and passed me a mug of tea, so I stopped work and sat on the boiler barrel in order to enjoy the welcome beverage. I should add it was a very hot boiler barrel.

Big mistake! As I sat there drinking my tea, the heat from the boiler warmed up that part of my boiler suit with which it had contact. A few minutes later, and completely without warning, there was a flash of bright flame as the solvent on the fabric ignited.

I have never taken a boiler suit off so quickly, but unfortunately not quickly enough - hence the name 'Nut Roast'. As I finished telling Fred the tale he started to walk away, but not before exclaiming loudly the title of a 1960s hit song by Jerry Lee Lewis! He never let me forget that incident.

He was a great inspiration to a lot of people, in the steam job in particular and vintage machinery restoration in general, and he raised the level of interest in the subjects by his brilliant TV programmes.

He was an icon for the movement, on which he has certainly left his mark and I, along and with many others, certainly feel the better for having known him.

From living van to Buckingham Palace, all in a days work for Fred. As Alf gets busy preparing to move off Fred tries on his 'zoot suit'.

On the Road with Fred
Alan Atkinson

Atkinson Trailer Hire is a Lancashire company which operate vehicles that many will have seen zipping up and down the country's motorways, but those given to attending vintage shows and steam rallies around the UK will have also noticed the firm's low loader. Inevitably, it would be delivering or collecting someone's cherished steamer to a show site. Consequently, when Fred was asked about a firm that might possibly help moving the convertible between filming locations, there was only one name he gave to the film-makers, that of his long-time friend, Alan Atkinson. This is his story:

It must be every person's ambition to make a living from their hobby. For several months last year I was able to achieve my ambition by combining my love of steam with my job as a haulier. I was asked to transport Fred's recently-restored Aveling convertible tractor around the UK for what was to become the last BBC series, and I jumped at the chance.

My own involvement with the 'steam job' was via a baptism of fire that dates back to 1975. My brother Richard had helped in the restoration of a very desirable 1925 Wallis & Steevens 'Advance' roller. Unfortunately Richard was killed in an industrial accident shortly after the engine had been restored and I was thrown in at the deep end, so to speak, when I was asked to take his place in completing the restoration and transporting the engine – two things I had not previously been involved with.

Fred and Jack load the Aveling on Alan Atkinson's low loader.

Almost ready for the off, the boys check that everything in the living van is OK as Fred and Alan pose for the camera.

Alan's mobile café is open and two customers await their early morning cuppa.

Fred, with a brew, supervises Alan Atkinson and Jack as they lash on the convertible.

The owner had touchingly named the roller 'Friend Richard' in memory of my late brother.

It may have been meant to be – I shall never know – but it happened that the roller came up for sale in 1980 and I bought it and have thoroughly enjoyed rallying it regularly ever since. What is for sure is that I had caught the steam bug as I went on to add a half-scale Atkinson 'Colonial' steam wagon to my collection some years later.

I still take off with the family, given the slightest chance, to rally the two machines around the country. Summer weekends are, of course, all the better if there is some steam involved.

Over the years I had built up a good relationship with Fred, even transporting him and his Aveling roller to the odd rally and promotional event up and down the country. A few years ago, as Fred's second engine was nearing completion, he told me he would like to make a film visiting important places throughout the UK and that he was going to talk to producer David Hall about it.

I told him to put my name down for the transport, never thinking that one day a telephone call from a TV company would answer my prayers. The highlight of the whole affair was the trip to Buckingham Palace when Fred got his 'gong'. We couldn't park up on The Mall but did stop a little way from the front of the big house; we unloaded in Wellington Barracks on Birdcage Walk.

The 'secret of my service' while filming was really very simple, and to those who knew, no secret at all. I always kept a portable gas stove, kettle, water and brew requisites in the cab. While the TV people got set up for the day ahead film-wise, my most important job after lighting up and unloading the engine was to brew up.

Fred would arrive, usually with Alf and Jimmy, and head straight to the low loader. Picnic chair out and brew in hand, he would then hold court with his public.

As the summer wore on, the whole crew got to know where the brew gear was stored and I ended the location filming with a total of eight beakers lodged in the cab. Oh, yes, and the odd tin of baked beans.

One of the great pleasures during last summer was being able to witness first-hand the affection between Fred and his two fine boys. Jack and Roger spent many hours both on location and on the road with us and they were very much a part of the team. Whether it was by breaking up wood to light the fire, cleaning the engine, fueling up the coal bunker or getting to drive and fire the Aveling, they were constantly involved.

The fellowship we enjoyed in the evenings after filming will also not be forgotten. Fred was a well-loved figure and, being in his company, we were all treated like royalty wherever we travelled. It was sad to see Fred's health deteriorate throughout the tour but, however unwell he felt, he always had time for his fans.

'On the Buses' to 'Up the Ladders'
Neil Carney

Neil Carney has spent the majority of his adult life as an engineer working on public service vehicles and he took early retirement from his final job as depot engineer at the Wigan bus garage of Greater Manchester Transport to concentrate on his hobby of engineering model making.

Mr Carney is a very accomplished modeller and, in addition, shares a love of all things steam with his wife, Frances. This passion brought them into direct contact with Fred Dibnah who, by all accounts, was already aware of Neil's engineering skills.

From those early meetings the two men developed a comradeship that eventually led to a second career for Neil as a steam restorer and steeplejack's 'grounder'. In the fullness of time he even ventured up to the top of a 'quite big 'un' and, as the men's work involvement grew, so did their friendship. The Carneys' daughter, Diane, went on to become godmother to Fred's youngest son, Roger, and even after Neil stopped working for Fred he kept in regular contact. Neil recalls that they were, in fact, lodging away while working on their first restoration job together on the day that Roger was born, 4 June 1990. This is Neil's story:

I first met Fred about 1966 at an early Burtonwood traction engine rally, which was about the time he had just acquired his own roller. At that time we were living and working in our home town of Ormskirk but, in 1969, because of my work, we moved to Bolton. I therefore lost no time in visiting the Dibnah 'works' to check on the progress of the roller I knew Fred was rebuilding.

I recall that he was working on it outside in fairly basic conditions at that time. In fact, if I remember correctly, all he had was a tarpaulin sheet slung between two trees. We kept in touch, but I didn't see a great deal of him until 1989, as it was then that I had taken early retirement and called at the yard to tell him so. Fred lost no time in asking if I might be interested in assisting him with the re-erection of a steam engine belonging to a mill in Wales. This he had removed during the winter of 1988-89 and, having refurbished it in Bolton, it now needed taking back to Caernarfon and reinstalling.

The site was at Glynllifon Park near the village of Llandwrog, and we worked on the project on and off for about nine months. In addition to the engine, we fitted a new boiler rescued from a meat pie factory. At the end of the job the customer had a refurbished engine that worked perfectly. That was towards the end of 1990. I think Fred appreciated my engineering skills and I was able to put them to work effectively during this installation.

We had worked well together, and I think that was why Fred asked if I might be interested in staying with him on the steeplejack side. I was keen to accept, although, I must add, I insisted that the deal would not include climbing and I would only be his 'grounder'. That was to be the beginning of five happy years together, where about half the time was spent on chargeable work, and the rest spent, as I would put it, 'playing in the garden'. Those who've visited there will know fully what I mean. There was always some engineering project to tinker with and I thoroughly enjoyed myself.

A fair amount of the time spent at the yard was dedicated to improving the line shafting and drives associated with Fred's many pieces of machinery. This included connecting a

Neil assisting Fred with the flywheel on the convertible.

Fred and Neil with the engine reinstalled at Glynllifon Park.

new section of shafting at right-angles to the existing one and driving it by way of a gearbox. I mention this because Fred initially used the gearbox (connected to a belt and pulley) to drive all his machines by using the steam roller. Eventually, of course, the machines were driven by a dedicated steam engine, a unit which had originally been employed to power a mechanical stoker associated with a Lancashire boiler at a cotton mill.

With regard to chimney work, there was not always a lot for me to do while the actual repair work was taking place 'up top' and it seemed then that a great deal of Fred's time was taken up by pointing. He liked to use special mastic, which he mixed for himself using red lead. This, he maintained, resulted in a better quality job and was in the event easier to use.

Laddering was the real hard graft. The sections would all be prepared a day or two before (ie tying the 'skids' to the ladders). These are the little goalpost-like brackets slightly wider than the ladder, which ensured that it 'stood off' some nine inches from the chimney or tower. The necessary number of ladders was then loaded on the trusty Land-Rover together with 'the dogs'. These were the fixing spikes intended to be driven into a wooden peg, which had first been knocked firmly into a chiselled hole made in the brickwork of the chimney.

To complete the loading, two lashings per ladder (ie a five-foot length of rope with a tied loop at each end), were also checked on to the vehicle. These sections of rope are required to tie the sections of ladder to the dogs, which will have been fixed to the chimney. A rope to a length totalling a little over twice that of the height of the chimney and a pulley block to be used in conjunction with it would then be added to the cargo.

The full-size rope is then used by the groundman to haul each ladder up to the man at the very top of the previously fixed ladder. In this manner, the chimney is fixed and provides a means by which the steeplejack can safely reach the area intended for his work. On a good day, we could erect up to five ladders in one hour. If the chimney was to be decked out, then it was usual to fix two ladders against it, on opposite sides.

The two tallest chimneys that I worked on with Fred were respectively at India Mills in Darwin and Barrow Bridge in Bolton, the latter being 262ft high. On the Barrow Bridge chimney we used two runs of ladder with eighteen sections in each and a section of decking around the top. The contract called for the pointing of the top 30ft and then recapping the top brickwork.

On the other hand, the India Mills chimney was laddered so some guy could abseil down it to raise money for charity. He insisted on a rope being tied round his waist as he climbed to the top on the ladders (that, in turn, being around a pulley and held firm by someone, as he admitted to being 'not happy' on the ladders). He had abseiled from balloon baskets and was perfectly happy once he got off the ladders, and I recall that he succeeded in raising the necessary cash for the charity concerned.

I did eventually work at height, but not on a massive chimney. It was a water-wheel tower near Todmorden. It was a very impressive stone structure some 70ft high, which had originally housed three waterwheels, one above the other. Our job involved reconditioning the capstones around the edge of the tower, which was about 30ft by 12ft in girth.

Fred had erected a walkway inside the tower about three feet down from the top and I was quite happy working on that. We were renewing the metal straps that tied the stones together and, as the job progressed, the tools we were using got scattered around the job. To make things easier, by his way of thinking, Fred decided to place a couple of extra planks across the middle from one side of the inside of the tower to the other.

I said to Fred: 'If you want me to walk across there then let's have one of those spare scaffold poles fixed as a handrail.' After all, we were 70ft above the ground. In his usual way, he just said: 'No problem' and set about fixing one in place.

After that I just got on with the job, never bothered about the 70ft drop and happily shuffled back and forth across that plank, because the improvised rail was there. Fred must have noticed this and he made no comment until the end of the job. 'Don't know why you wanted that rail, Neil. You never used it once,' he observed. And he was quite right. I hadn't.

I was always very busy on the chimney-dropping jobs as my major task entailed keeping the work area free of broken bricks as the boys on the chisels or jackhammers cut them out of the chimney. It was important to have a level working area to make it safe while inserting the props. These were telegraph poles, cut to the exact required length using a chainsaw. The length varied slightly, depending on the thickness of the piece of wood used as a cap piece, which was fashioned to bridge the inner and outer props above the wedges. The wedges were then driven in as pairs, as tight as possible, by using heavy hammers and a lot of muscle power.

To give you some idea of the work involved, the wall thickness of a tall chimney could be up to four-feet, and to drop a chimney like that we would be required to use approximately 16 props and remove 180ft of brickwork. As Fred developed his technique,

he took to drilling the support props to aid the burning process. He originally achieved that by using a hand-powered breast drill though, as time progressed, he used power tools for the task.

Fred never considered any alternative to this method of downing chimneys and he never liked or trusted explosives. 'You can never be sure where those few bricks which fly out sideways are going. With this method I have a fair idea and total control – well, more or less!'

If I had to sum up Fred's abilities, I think there is one word that covers almost everything and that word is 'ingenuity'. He was a good friend, great guy to work with, a natural engineer and a quick thinker.

One story, in particular, illustrates the way his mind worked. He was returning from a vintage show, miles from anywhere, together with a pal, when the Land-Rover got a back-wheel puncture, and not a pub in sight!

The pair jumped out and examined the problem: it was flat all right. Fred had a spare wheel and its tyre was fully inflated, but what he didn't have was a jack of any kind. That was, except for a 12ft length of 4in x 2in timber, which he had bought at the show to use on a job later in the week.

A cheeky grin as Fred prepares to ride off into the night, something he greatly enjoyed doing.

His thought processes went something like this: "We can lift the vehicle using the timber as a lever and hold it high enough, for long enough, to be able to change the wheel. His puzzled companion agreed with the plan, but asked Fred what he planned to use as a fulcrum?

"No problem," said Fred and, grabbing a spanner from the back of his vehicle, he then forced up a cast-iron grid from the side of the road. We are all familiar with the phrase 'that'll do nicely'. Well it did!

Fred Said!

Did yer like that?

The modern world stinks.

I'm just a bum who climbs chimneys.

A man who says he feels no fear is either a fool or a liar.

One mistake up here, and it's half a day out with the undertaker.

We've become a nation of con-men, living by selling double-glazing to each other.

Height gives you a wonderful feeling of grandeur. You're the king of the castle up here.

I realise that steam engines aren't everyone's cup of tea, but they're what made England great.

Steam engines don't answer back. You can belt them with a hammer and they say nowt.

I set out as a steeplejack in my youth to preserve chimneys. I've finished by knocking most of them down.

Anybody who destroys anything made of stone should be prosecuted. It is not all beautiful, but it took a man all day to cut one stone.

The Day that Bolton Stood Still

Funerals by their very nature are dour, grey events and, try as they may, the clergy rarely succeed in entirely convincing the assembled mourners that the event is, in fact, a celebration of the dear departed one's life on earth. The occasion of Fred Dibnah's last journey was, without doubt, the exception that proved the rule.

He was an iconic figure to many, both at home and overseas, and he will be greatly missed. To his family he was, of course, much more: a husband, a brother, a father, a grandfather, a cousin, a relative, and it is for those people that his loss will be greatest. He was their kith and kin whom they unselfishly shared, first with his close friends and then with the many who knew and admired him. His funeral day was not an occasion when the family could hide away and mourn in private. It was a very public affair. The family shared their day of grief with so many.

The early morning gloom which, on Tuesday, 16 November 2004, pervaded central Lancashire, was to grow by the hour, until it developed into a traditional British rainy day as the poignant events unfolded.

And yet, as Fred's cortege set off from his home at 11.15am and headed for Bolton's Parish Church, the damp and muted conditions seemed appropriate.

From an early hour, many thousands of citizens from that famous old industrial town had lined the route of the funeral procession, where a great many others joined them who had travelled from far and wide to pay their last respects to this glorious son of steam.

Many of the faces of the hushed waiting throng were dampened as their own tears of sadness mingled with the steadily falling rain. The elements had, in their own way, provided a fitting backdrop to the last public act of that town's greatly-loved favourite son.

Fred had requested a Victorian-type funeral – but with steam. This led undertaker John Howarth to comment, when asked by the family to officiate, 'Organising this very important event has been a first for us, but I think it will do him proud.'

On the behalf of those who were privileged to attend and, indeed, those millions more who saw the event on TV, may we say that you, Mr Howarth, did exactly that. Some who knew Fred have commented since that, on the day, it was almost as if he were directing the proceedings himself.

Many, through recent contact with Fred and perhaps from their own family experiences, were aware of the ravages of cancer and there was a palpable air of relief among the mourners. A dearly loved friend was no longer a sufferer from a disease that, in one or another of its many forms, touches one-third of the population. He was free of earthly pain and now deservedly at rest.

Fred's life touched many people in differing ways and people worldwide have rightly noted and talked of his passing with sorrow. To the townsfolk of Bolton it was so much more, and deeper: they had lost one of their own. Moreover, in this multicultural part of the Red Rose County, Fred was a 'mate' to people of all colours and creeds and the whole community mourned his passing with equal admiration and respect.

Leading the cortege was the band of 103 Regiment, The Royal Artillery 'Bolton Volunteers' from the town's Nelson Street Barracks. The smartly-dressed, but accordingly sombre musicians were followed by Fred's long-standing pride and joy, his restored steam

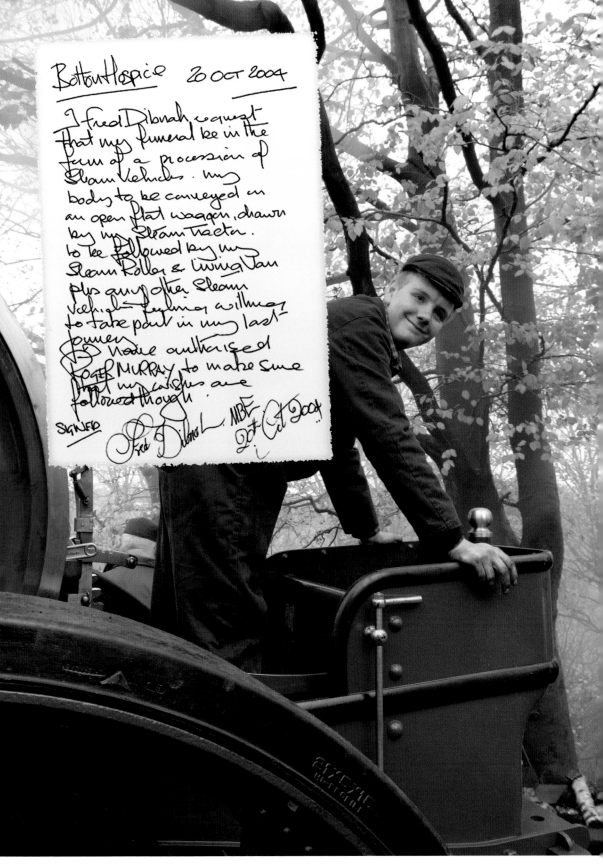

A proud moment for a brave young man. Fred's elder son, Jack, checks that enough coal is on board the convertible before helping to crew the engine, which is to pull the trailer conveying his dad's coffin.

A family floral tribute in the hearse, as Fred arrives to take his last journey hauled by steam power.

The waiting crowd fall quiet as Fred arrives outside his Bolton home.

The band of 103 Regiment, The Royal Artillery 'Bolton Volunteers' waits to lead the cortege.

The huge crowd outside Bolton Parish Church, having waited patiently and silently in the rain, are stirred as the cortege enters Churchgate.

The band entering the square in front of the church.

The band comes to attention awaiting the coffin.

roller 'Betsy'. Crewing the roller were two of Fred's lifelong pals, Roger Murray and Neil Carney, with Roger, Fred's youngest son, in close attendance.

The living van, so lovingly restored by Fred, was in tow with his faithful old Land-Rover attached. Travelling on the steps of the living van were two friends who additionally had journeyed with Fred to various filming locations during 2004, ex-miners Alf Molineux and Jimmy Crooks. Dear friends and colleagues too numerous to mention individually had gathered on that rainy day in order to say their goodbyes, the townsfolk and their visitors were to a person in deep mourning.

The newly restored Aveling & Porter and its crew had perhaps the greatest responsibility of all as it was chosen to draw the flat trailer on which Fred was to ride. It is to the youthful steersman of the convertible that much praise for a job well done, under great pressure, must go. Teenager Jack Dibnah did his father, his family and Fred's admirers proud, not only handling the engine with skill but also presenting himself with the utmost dignity.

Michael Webber, a native Boltonian who now resides on the South Coast, yet another of Fred's many friends, was Jack's partner and it was he who had supplied the flat truck. Fixed upon it were a selection of Fred's steeplejack ladders and ropes. Michael had also secured brackets to hold the coffin firmly in place.

Dave Guest from BBC North West filming his piece from outside the church.

Fred's beloved steam roller 'Betsy' hauls his living van, which in turn is coupled to his Land-Rover, into Bolton's Churchgate. The roller is crewed by two of Fred's friends, Nigel Carney and Roger Murray, and Fred's younger son, Roger. Roger Murray was deeply involved with the organisation of the funeral, having been a confidant and aware of Fred's last wishes.

The two friends who accompanied Fred on his latest filming escapades fittingly ride on the living van steps. Nearest the camera is Jimmy Crooks and, on his left, Alf Molineux.

Fred's elder son, Jack, steers the engine through the waiting crowd and into Churchgate.

Aboard the convertible with Jack is a long-standing family friend Michael Webber. A Boltonian who now resides in Worthing, Roger provided and 'rigged' the trailer on which Fred's casket is carried, along with items of his steeplejack gear.

A model of the adopted Bolton symbol, an elephant with a castellated howdah, on loan from the local museum, was fixed at the head of the casket on which Fred's famous flat cap rode forlornly for all to see. Four other steamers, representing Fred's many acquaintances and admirers, followed closely behind.

As the immaculate and perfectly poised, John Howarth led the cortege into the town's Churchgate, the waiting host who, until that point had stood in solemn silence, began to stir. Gradually they caught sight of the vehicle carrying Fred's earthly remains. The silence was broken by, at first, a small ripple of hand clapping, which then rapidly grew into a veritable crescendo and then ceased just as quickly, as the customised 'hearse' came to rest outside the church. The people of Bolton had, in that spontaneous act respectfully said 'thank you' and bade goodbye to Fred.

In truth, double the seats would not have been enough to accommodate all those who wished to partake in the service of remembrance.

The crowded church, a building that in years gone by had benefited from Fred's steeplejack skills, fell silent as his coffin entered to the strains of Julie Covington's rendition of *Don't Cry For Me, Argentina*, a favourite piece of music of Fred's.

For many trying to hold on to their emotions the welcoming words of Canon Michael Williams, Vicar of Bolton, were a timely consolation. How proud Fred would have been to hear his daughter, Lorna, read perfectly the poignant words of Rabindranath Tagore's poem *Farewell My Friends*, a fitting tribute delivered with tenderness and love. The congregation cleared their throats and, with gusto and feeling, sang *The Lord's My Shepherd*; the vaulted roof of this impressive Lancastrian edifice seemed to echo the fine words of the 23rd Psalm.

The Rev Barry Newth had been associated with Fred and his family for many years and, indeed, was instrumental in awarding a bell tower repair job to him in the early days. The two men remained friends and, fittingly, Barry was at Fred's bedside, in the company of family members, when the end came. Some days before, Roger Murray had visited Fred in Bolton Hospice and, on the bedside table, had discovered a piece of paper on which, in Fred's distinctive hand, was written simply the name Barry Newth. Roger knew that Fred wanted to see again his dear friend.

Barry enthralled and amused the congregation with many fitting tales of Fred's exploits, not least of which recounting the time he had to resort to climbing. Fred had persuaded the terrified vicar to accompany him out on the roof of his church and then coined a one-

Note Fred's 'trademark' flat cap on the top of his coffin. Also aboard the trailer is an elephant which has, over time, become associated with the town of Bolton and dates from the time when the town was part of the diocese of Coventry, it being a symbol also associated with that city.

liner, which later became part of the Dibnah folklore. Fred advised Barry thus: 'Be careful vicar. One false move and it's half a day out with the undertaker.'

He told of the time Fred turned up as promised for a church fete with his beloved steam roller, but very late and a little the worse for drink. He intimated that he naturally got Fred to pay for that misdemeanour with good deeds many times over the years.

David Hall, who for many years had been Fred's film producer and co-author, continued with anecdotal tales from the other side of the camera. It was obvious to all that the two men had become a lot more than just work colleagues – they were friends. Indeed, as he explained, so natural a performer was Fred that the whole production crew benefited from their experiences with him. The problem working with Fred was, said David: 'Trying to keep to a schedule which was, to say the least, difficult, given that he would stop to sign every autograph, pose for every photograph, and shake every hand.' That was Fred.

Roger Murray was the next to address the assembly and, having arrived in church fresh from the footplate of 'Betsy', wearing his 'blues', there was perhaps no more fitting a way to honour Fred's memory and represent engine men and women everywhere. He spoke touchingly about his visits to Fred in the hospice and, in so doing, shared with all the sadness of those last days. He told of how Fred asked him if he believed in Heaven. And of how they went on to talk of Fred possibly meeting his Victorian heroes and others who had gone before.

At the conclusion of the service, and following the commendation and ministration of Canon Williams, Fred's coffin was taken from the church to the strains of Carnival of Venice (the theme from his TV series) and repositioned on the steam-drawn hearse for the journey to Tonge Cemetery. No greater compliment could have been paid to his memory as the heavily falling rain in no way prevented huge crowds from lining the route of his final journey.

At Fred's wishes the attendant engines sounded their whistles in unison as he was lowered to his final resting-place.

The Words of Barry Newth

One of the closest of Fred's friends was the Reverend Barry Newth and it is appropriate after Fred's untimely death that we remember him with some of Barry's thoughts:

'How can I ever forget my good friend Fred Dibnah – and in any case, why would I, or indeed anyone else, want to?

'I first made his acquaintance in 1963, the same year that I became Vicar of Clifton, a village a few miles from Bolton on the A666 to Manchester. The one bell on my church had broken away from the open bell frame on the church roof and needed urgent attention.

To whom could I turn?

As a complete stranger to the area I was directed to Canon Greville Norburn, then Vicar of Bolton. He'd just employed a young man by the name of Fred Dibnah to repair the weather vanes on the church roof.

'He's done a good job,' said the canon. 'Try him out for yourself.'

From that first meeting in 1963, Fred and I shared forty-one years of friendship – not that we saw very much of each other, in fact, very little for many years. Fred was far too busy answering the calls from local vicars and factory owners. With the latter he eventually began the onerous task of felling chimneys. It was this, and more, which gave Fred a nationwide audience on TV.

'So be it'

The crew of 'Old Faithful' getting a soaking but still up to the task as they bring up the rear of the cortege that wends its way through the town and on to the cemetery.

'Betsy' outside Fred's home on Radcliffe Road.

The convertible on the way to Fred's last resting-place in Tonge Cemetery.

Fred relaxing with his Aveling & Porter convertible steam traction engine at the top of Llanberis pass, North Wales.

The headgear is now silent, Fred's pit head gear pictured on the third anniversary of his passing.

Fred looking happy the day that his mine shaft headgear was finished.

A few years later he returned first with son Jack, and then, three years on, with Roger, for me to christen them at the Norman font in Kirkby Malham.

And what a privilege it was for me to take a small part in the first films Fred made for television, including the placing of a new weathercock on the flagpole. It was made from an old copper boiler!

During his short stay at the vicarage, my wife and I used to wonder what he did with his beloved flat cap. It never left him. He even took it to bed with him.

My last and most beautiful memory of Fred comes from my visit to Bolton hospice on 6 November, 2004. He'd even asked to see me and I responded immediately to the telephone call.

I sat by his side with his family. He was only just conscious. I held his hand and recalled with him all the good times we had together. I'm sure he recognised my voice. And just as I said a final prayer for him and we all started to say the Lord's Prayer together, he breathed his last.

Why would we possibly forget you? Fred Dibnah was a lovely man and a real character and he possessed great courage and perseverance, both wonderful marks of his Christian qualities.

The Making of Fred...

Shropshire is steeped in industrial traditions, and accordingly Fred Dibnah would have been pleased that the sculpted clay for his bronze effigy was created in Shrewsbury. Furthermore the actual casting was undertaken in a superbly rural setting just to the west of the once famous railway town of Oswestry. To be accurate, the bronze foundry of Castle Fine Arts Ltd is actually in the principality of Wales – perhaps the village name is a bit of a giveaway – Llanrhaeadr ym Mochnant (it's in Powys).

Bronze has been used to cast sculpture since ancient times. Today's foundries employ some modern techniques and materials but the process is basically the same one that man has been using for 3000 years. If a foundryman from ancient Greece or Rome could visit Castle Fine Arts today he would be familiar with many aspects of the production process.

In order to create our Fred's 'Bolton Bronze' an amalgamation of two techniques were employed; 'lost wax' from times past coupled with the more modern process of 'ceramic shell' casting. The revered Rolls-Royce 'Flying Lady' emblems were created using the same technique, and so it was a fitting process by which to make the Fred statue. Before the foundry can start their work, the item to be cast has to be made in a material from which the mould(s) can be taken. That was the task set for sculptress Jane Robbins.

Miss Robbins was chosen from a shortlist of several sculptors. She trained at London Guildhall School of Music and Drama in theatre arts and in figurative sculpture at Stafford College. She has produced a highly detailed memorial bronze plaque of HRH Queen Elizabeth The Queen Mother as well as the life-size bronze of Linda McCartney, which is situated on the Mull of Kintyre in Scotland. Another of her public works of art is the Andy Capp figure, which was made for Hartlepool – home of the famous cartoon character's creator Reg Smythe. Jane has also produced a bust of the late great Salopian gardening expert Percy Thrower.

The Civic Trust ordered the statue to be 'over life size' at 8ft high as their original intention was to place Fred on the top of a 6ft tall replica chimney top, with a lightning conductor in hand. However, the planners did not like that idea and insisted on a modest 'just above ground level' plinth instead. That fact accounts for the statue depicting Fred in a slightly forward leaning pose, as he would have been if looking from the top of a chimney. On ground level Fred would of course have adopted his customary 'tummy out shoulders back' stance!

Firstly Jane Robbins fabricated a welded and shaped steel frame (armature) on which to place the clay. In forming the statue she used more than a quarter of a tonne of the specialist modelling material and spent over six weeks carefully crafting Fred's features. She even managed to capture his famous grin, his trademark flat cap, his gold watch chain and his spectacles. When Jane was happy with her work, and after it had been given the green light by the Bolton Civic Trust, she arranged for the foundry technicians to start their work. Because of the sheer size of the 'piece' the founders visited her English Bridge Studio in Shrewsbury, in preference to shipping the 'clay' to the foundry in order to carry out the next part of the fascinating process.

Castle Fine Arts is home to one of the UK's leading sculpture foundries, with studios in North Wales, Gloucestershire and Shropshire. They work closely with a wide variety of artists to create large-scale bronze sculptures for public and private commissions, both in

Sculptress Jane Robbins pictured in Bolton with the finished work. She used more than a quarter of a tonne of specialist modelling material in creating the 'Dibnah Bronze' and spent over six weeks doing so.

Work in progress at Jane Robbins's English Bridge Shrewsbury studio.

A wax mould seen after being stripped from the rubber coated glass fibre holder.

A large number of moulds are seen drying after being coated with layers of ceramic slip and grit. The various 'runners' and 'risers', which will allow the molten bronze to fill the mould, and let the air come out, and the funnel shaped 'cup' can all be seen.

the UK and overseas. It is a specialised field in which the firm has gained a wealth of knowledge. In addition to actually producing the artworks they provide a range of technical and practical support to sculptors including consultancy, project management and a commissioning service.

The term 'lost-wax process' is so called because it requires creating an exact replica of the original sculpture in wax. Lost-wax is perfect for casting sculpture because any shape can be cast and fantastic 'finger print' surface detail from the original sculpture can be reproduced. The advantages of the lost wax and ceramic shell process were recognised early in the industrial revolution by manufacturing industries and in modern times turbine blades for jet engines have been made this way, thus having the advantage of requiring very little machining.

The first stage of the casting process is to make a mould in silicon rubber. The flexible rubber creates a negative version of the sculpture. In effect a container into which the molten wax can be poured, in the case of Fred's statue the wax was built up to a thickness of approximately a quarter of an inch (which represents the wall thickness of the hollow statue). Because of the sheer size it was cast in ten separate panels plus head, hand and lightening conductor. The 'negative' moulds were then encased in glass fibre to give them rigidity.

The hollow wax (held firm by the glass fibre) is a replica of the original sculpture and all of the mould seams need to be skilfully worked in order to clean up the negative surface which will become the positive (outer) surface of the casting and thus represent the intentions of the artist. Art foundries have always formed close working relationships with sculptors and the artist will often come to the foundry to check it before casting. It's a chance to make any minor changes as the next time they see their sculpture it will be in cast in bronze.

With the 'wax' created, another mould has then to be made, but this time in ceramic. On to the wax mould the foundry workers first stick wax bars called 'runners' and 'risers,' which will allow the molten bronze to fill the mould and let the air it replaces to escape. A funnel shaped 'cup' in which to pour the molten bronze is also added. The whole piece is then dip-coated in a ceramic slip material and layers of ceramic grit are applied.

Often this is done five or six times before the desired thickness of ceramic shell is achieved. The resultant work piece is then fired in a kiln at a very high temperature to achieve two things. Firstly to melt all the wax out (ie lost wax) and then to harden the ceramic to create a pot (in effect a double skin with a hollow centre). The ceramic shell work piece has to be strong enough to pour molten bronze into. The bronze is melted in the furnace to a temperature of over 1120 degrees centigrade and then carefully poured into the shell. In order to hold it still during the pouring process the work piece is immersed in a container of fine sand, on which weights are strategically placed.

It's an exciting time when, in the noise and the heat of the foundry, the sculpture is cast, the shell is then broken away and we get to see the results of the founders' efforts and the sculptor's skill. This really was the case as we witnessed the steaming hot ceramic shell of Fred's head lifted carefully from the sand and placed on the foundry floor by the three skilled men comprising the casting team. They do say that 'when Fred was made they broke the mould', in this case they literally did just that! As the ceramic shell was 'encouraged' to come away from the rapidly cooling bronze the first thing to be recognised was the peak of Fred's cap.

The whole piece is then dip coated in a ceramic slip material and layers of ceramic grit are applied. Often this is done five or six times before the desired thickness of ceramic shell is achieved.

Back in the Castle Fine Arts foundry, Scott Shaw is pictured applying a coating of wax to the inside of a section of a mould.

Production Manager Chris Weston with the glass fibre surround with rubber mould from which the wax casts of Fred's face were formed.

The crucible is filled with bronze and the temperature allowed to gradually build up.

The top of the lightning conductor as cast.

A place in the country! Castle Fine Arts Foundry at Llanrhaeadr ym Mochnant.

Two work pieces, Fred's head and right hand are placed side by side in the sand box in order to be to be poured at the same time.

The temperature of the molten bronze reaches 1100 degrees centigrade. The pour will take place at 1120 degrees C.

The pouring process.

On pouring duty the day of the OG visit were (from left) Geronimo (Jez) Latto, Mike Annitt and Aaron Hines.

From Iceland with love; young lady art worker Gudlaug Jensdettir gets to work fettling Fred's hand. He would have admired her skills!

The filled mould is gently lifted from the sand box, a moment of real anticipation!

The statue begins to take shape as the sections are welded together and fettled.

Fred's head as a ceramic shell, again the runners and risers can be seen, in this image the 'cup' into which the bronze will be poured is the base on which it temporarily stands.

Once cast only half of the work is done! Then different skills are needed to 'finish' the bronze and to recreate in metal all the forms and marks from the original sculpture. The large sections were welded together and the surfaces all 'worked' to lose any traces of the process. The metal workers at the foundry need to have a clear understanding of what the artist wanted to achieve, and furthermore possess the expertise to make it happen.

After welding all the sections together, Fred was then given a pair of 'specs' and the lightning conductor was added to the statue and welded into place. After each of the pieces was cast a certain amount of 'fettling' took place and then after welding all the pieces together the final finish was achieved, but only following a great deal of skilled work by the team. As a certain Mr Morecambe would have said, you can't see the join!

So there you have it – an 8ft statue of 'Our Fred'. Who would have thought it? Certainly not him!

Art, as they say, is subjective. It will be seen differently by almost everyone who looks at it. Those who knew Fred will probably expect to see the last image of that dear man which they hold in their memory, others will maybe recall Fred from a particular incident or TV scene. We will all see what we wish to see, but in doing so we will remember Fred.

Jez Latto encourages the ceramic coating to part company with the bronze.

The ceramic shell is gradually removed as the work piece cools down.

The cast of Fred's head awaits fettling.

A Bronze for Fred

The date 29 April 2008 would have been the late Fred Dibnah's seventieth birthday. Although a modest man, Fred would have loved the memorial to his memory unveiled in Oxford Street, Bolton, on the anniversary of that occasion, especially with his five children looking proudly on,

But the significant bronze statue was only made possible by the generosity of his many admirers and friends. Thousands of people nationwide contributed to the fund set up by Bolton & District Civic Trust, On 6 November 2004, the day Fred was sadly taken from us, his long-standing friend Bill Greenhalgh vowed that the town of Bolton would have a suitable memorial to its famous son. Unbeknown to Bill, a lady journalist on the other side of town was having similar thoughts!

Bill found a fellow traveller in the person of Brian Tetlow, chairman of the town's Civic Trust. The pair met with the local paper where deputy editor Lynn Ashwell told them of her ideas. 'Lots of people had statues in the towns where they were famous. Eric Morecambe's famous pose adorns Morecambe and is a tourist attraction. Footballer Billy Bremner stands guard at his beloved Leeds. So Bolton needs a Fred.' How right she was!

The scheme quickly got under way and an appeals committee was formed. Fred's public loved the idea, they picked up the baton and ran with it and the race was on.

Fred's very proud children, seen after the unveiling ceremony. From left: Jayne, Lorna, Caroline, Jack and Roger.

The grandchildren get a day they'll remember: Christopher and Daniel (Jayne's), Isobel (Lorna's) and Jack (Caroline's).

Fred, the man of the people, among the people of Bolton who turned out in force to witness the unveiling on what would have been Fred's seventieth birthday.

Fred holding a lightning conductor aloft.

Fred at one and a half times size in front of the Hick Hargreaves engine that he helped to save.

The ambitious plan was to first raise the money (over £40,000) and then commission and erect a bronze statue in time for what would have been Fred's 70th birthday – a tall order, with a timescale of just three and a half years.

But against all the odds, the race was won. The finishing line was crossed at just after noon on 29 April as the Mayor of Bolton, Coun Barbara Ronson, unveiled the statue.

Prior to the unveiling television producer/director David Hall made a short speech outlining the significance of the location of the statue and fondly recalling the time he had spent with Fred, confirming that the two men became not just media colleagues but good friends. After Brian Tetlow, of the Bolton Civic Trust had thanked his hard working committee members and the local council for their support it was the turn of Fred's widow Sheila to address those gathered to witness the event.

Mrs Dibnah thanked all who had worked so hard to raise the money which had made the creation of the statue a reality, telling the onlookers that the town centre location was a great choice as the preserved Hick Hargreaves Improved Corliss engine displayed nearby had been one of Fred's favourite restoration projects.

In a 2005 ceremony a prestigious Blue Plaque was unveiled on the front of 121 Radcliffe Road, Bolton, Fred's former home. The inscription on the plaque reads: 'Home of the late Dr Fred Dibnah, MBE, steeplejack. Honorary doctorate, Aberdeen and Birmingham universities, artist, draughtsman, carpenter, stone mason, demolition expert, intuitive engineer, steam enthusiast, devotee of our industrial heritage, raconteur and television celebrity. Revered son of Bolton 1938–2004.'

Fred, always a man of the people, now stands proud among his friends in the town he loved. The buzz among the crowd fell to a reverent hush before the cover was whipped off to unveil Fred's smiling face and people jostled to get a look at his bronze, 8ft-tall sculpture, one-and-a-half times Fred's actual size, designed by Jane Robbins. It depicts him in work clothes and trademark flat cap, holding a lightning conductor.

'Did yer like that?' The assembled crowd certainly did. Immediately after the unveiling, songwriter Pete Martin led the crowd in a rendition of Happy Birthday as well as his own original tribute song, *Ohh, Ohh Fred Dibnah*.

Fred stands alongside the encased Hick Hargreaves Corliss stationary steam engine – an historic piece of machinery which he greatly admired and had a hand in saving. Sculptress Jane Robbins was relieved to finally see the statue in place. 'I'm very happy because people have been saying very nice things about it and that's all I wanted – the people of Bolton to like it. I hope people grow to feel that it's theirs, to honour a man that they're very proud of.'

The family were all pleased with the outcome and issued this statement to *Old Glory* on behalf of the children: 'Jayne, Lorna, Caroline, Jack and Roger wish to thank the Civic Trust and The Bolton News for all the effort and work involved in making this happen today, on what would have been our father's birthday. While the main thanks have to go to everyone who donated money to the fund, particular thanks must go to Brian Tetlow (chairman Civic Trust) and especially to Bill Greenhalgh, a friend of dad for over forty years, without whose inspiration this statue would never have come to fruition'.

Songwriter Pete Martin led the crowd in a rendition of Happy Birthday followed by his own original tribute song, *Ohh, Ohh Fred Dibnah*.

Job done! A proud Bill Greenhalgh on
29 April after seeing the statue of his
lifelong friend unveiled.

Dr FRED DIBNAH M.B.E.
STEEPLEJACK

Honorary Doctorates:
Universities of
Robert Gorton, Aberdeen
5th July 2000
and Birmingham
19th July 2004

REVERED SON OF BOLTON
1938 - 2004

My brother! It was a proud day for Graham Dibnah when he witnessed the un-veiling of a statue to his elder brother, in their home town of Bolton.

Meet the Author
Keith Langston

Heritage transportation specialist feature writer and photographer Keith Langston's collection of Fred Dibnah pictures and stories were first published in various editions and special supplements connected with *Old Glory Magazine* during 2004 and 2005.

Having been privileged to travel with Fred during his last year of TV filming Keith Langston compiled a large selection of unique images. A great many of which, when added to pictures loaned by the Dibnah family and many of Fred's friends, were then published in two Mortons' Heritage Media special publications in 2005 and 2006. This commemorative compendium contains the author's personal selection of Fred Dibnah articles and pictures.

Based in mid-Cheshire, England, Keith has been associated with Morton's Media Group Ltd for many years, contributing news and feature material on a regular basis across a

wide range of heritage titles including, *Old Glory*, *Heritage Railway*, *Tractor*, and *Heritage Commercials*. Keith is well known for his railway material as the originator of the *British Steam* series of Bookazines, and he is also Chief Correspondent for the popular inland waterways newspaper *Towpath Talk*.